GUTSY

how the gut shapes the brain, the body and the woman you become

ALEXANDRA LIM, PH.D

PARTRIDGE

Library of Congress Control Number:		2026904675
ISBN:	Hardcover	978-1-5437-8569-2
	Softcover	978-1-5437-8568-5
	eBook	978-1-5437-8570-8

Print information available on the last page.

To order additional copies of this book, contact
Toll Free +65 3165 7531 (Singapore)
Toll Free +60 3 3099 4412 (Malaysia)
orders.singapore@partridgepublishing.com

www.partridgepublishing.com/singapore

CONTENTS

INTRODUCTION

There was a time not too long ago, when mornings greeted me with tears. Sadness and a sort of emptiness crept and settled into the gaps between bouts of work and meetings during graduate school. Even now as I type, an old ache is stirring, an ache that back then I couldn't quite place, because I did not yet have the language to understand how deeply my body was involved. On the surface, my life looked ideal. I had a stable home, a fiancé, and supportive friends. I was pursuing an exciting Ph.D., exercising regularly, staying social, and eating beautifully—anyone who knows me knows how much I love food. But behind that curated life was something messier.

Rewind many years. In my teens, I wrestled with a restrictive eating pattern that I half-recognised but never really named. It wasn't until my mother threatened to stop me from studying abroad unless I gained weight, half-pressured by relatives commenting on how gaunt and ill I looked, that I realised how visible it had become. I have a vivid memory of her eyes heavy with tears, hopeless and confused. I grew up with the privilege of abundance, yet I willingly chose to shrink myself down to nothing. For what? Still, for the longest time, I denied there was a problem at all, for I genuinely did not believe an issue existed. I later learnt that the inner workings of my gut in such an undernourished state had a large role to play in this belief system, directed by this interplay between my cellular physiology and brain health, at the time.

It worked in the short term, but the psychological foundations of a healthy relationship with food were not repaired by urgency or excess. I ate out of desperation, just so I could hop on a plane to London, where I was to pursue my undergraduate studies. The outcome was functional, but the underlying problem remained unresolved. I~~t worked, but the psychological underpinnings of a healthy relationship with food weren't healed by stuffing myself five to six times a day.~~

I wanted to prove that I was well enough to study abroad, to convince my mother and friends that I was just fine. So I did a total 180 and overcompensated, feeding a body that had been starved for years. I was hungry more for permission to get through a life I was determined to live and see through, not just for food itself. Food slowly became symbolic of opportunities I could expose myself to as life went on, not just something to look forward to three times a day. For the sake of efficiency, convenience became my comfort. My vegetarian diet at the time collapsed into piles of processed food, which was quick, easy, and numbing. My body swelled with it, yet I still felt hollow. I was bloated, uncomfortable, foggy, and emotionally dulled. I ate out constantly, no longer caring what I was putting into my body, and in time, my mind and spirit began to reflect that physical neglect. By the time I reached graduate school, the control I'd once clung to had slipped completely, turning into something else altogether. I no longer felt in charge of my body or what I put into it. I gained almost 20 pounds, swallowed whole by constant bloat and inflammation.

I moved through the days on autopilot, always busy, always doing something, for fear of stillness taking over and reviving that old ache in my chest again. I ate a lot then hardly ate at all, my mind and body always on high alert, never really knowing what they were going to face next. I hid the disarray behind my work and my social calendar, losing myself in research deadlines and late-night conversations that blurred into white noise. Beneath the surface of ambition and activity, an emptiness grew slowly but very surely, an emptiness that was quiet but all-consuming.

I equated control with confidence, in the beginning. It wasn't just about physical appearance; it was about mastery. I wanted to prove that I could outsmart my biology—that I could rise above my own appetite. At the time, I couldn't see how restriction and the subsequent loss of control around food were dimming me, dulling both my mind and my mood. My body wasn't the enemy; it was simply starved of what it needed to keep me grounded and alive. I've since learned, through years of education and experimentation, and as you will throughout the rest of this booklet, that lightness comes from *balance*, above and beyond everything else. From understanding the body, with particular attention to the gut, which quietly sits at the root of far more than we are taught to notice.

I kept wondering: *How does it feel to not have aesthetic ambition and inner equilibrium constantly in conflict?*

Or could the pursuit of feeling good in my body coexist with feeling good in my mind? For years, I thought restriction was the answer. That lightness meant less: less food, less indulgence, less pleasure, less richness. If I could just be smaller, more in control, then maybe the fog would lift, maybe the anxiety would quiet down, maybe I'd finally feel right in my skin. But the reality was otherwise. Ironically, the more I restricted, the heavier I felt—not in body, but in spirit and mind.

When you're stuck in that cycle, it's easy to confuse control for self-care. Restriction masquerades as discipline, when really it is deprivation wearing a slightly more virtuous mask. It's really quite sad that it had to take such a degree of suffering—physical, emotional, academic even—to understand that the human body never thrives under punishment, but nourishment. True vitality, that feeling of being clear-headed, energetic, confident, and yes, beautiful, comes not from cutting things out, but from knowing what to give yourself more of.

It took a long time, years, to understand that beauty and health are not opposing forces. They are partners. That food, movement, rest, and joy can coexist without guilt. I began to see my body not as something to fix, but as something to listen to. So when I finally began feeding it, combined with a newfound passion for gut health specifically, my mind started to organise itself around a new and exciting possibility: that I could build a life from clarity rather than punishment. And from there, the physical followed: an acutely strong, lithe, more attuned body, and it felt like every muscle fibre, neuron, and red blood cell of my body was working to keep a keen ear out for the inner workings of this sacred body, to work at its most optimised level, nudging me towards an expansive state of being.

Gutsy started as an exploration of that question and figuring what living more, as a healthy woman, means. Above all, through personal anecdote and research I was involved in, one thing is unmistakably clear: the *gut*—that overlooked organ we spend our lives ignoring or trying to flatten—holds the key to not just our physical health and optimisation, but our emotional and mental freedom. Through the gut, we understand ourselves very differently. We are not just these bodies to be controlled, but whole ecosystems to be supported. Once you learn how to work with this ecosystem instead of against it, you enter the very heart of this book's main thesis: physical optimisation and emotional freedom are not separate journeys, and becoming the healthiest, most confident version of yourself starts from one essential reset—learning how to listen to your gut.

Oxford was where this understanding began to crystallise.

Those years in graduate school provided quite a surreal backdrop to this period of my life, and to my own unravelling. As a student, you get to attend a bunch of formal hall dinners—traditional affairs where you meet eccentric folk over a mix of (sometimes uninspiring) wine and food. Sometimes I was invited to the high table, where the important people were seated. Not because I was particularly important—rather, people knew I'd slide in nicely. I enjoyed the intellectual exchange and had a knack for adapting my energy to all sorts of people. There was something enriching about exchanging ideas with others while feeling the energy of the room, the weight and register of its air, and matching my own to it. I'd learned to shift gears as and when needed, a skill that helped me navigate these events with ease.

As time passed, however, I felt both my social and intellectual enthusiasm wane during these dinners, and many a time, I simply wanted to disappear after the appetisers. Beneath the elegant dresses and wide smiles, I was entangled in a web of unhappiness and feeling uncomfortable in my skin, and really felt the weight of it—feeling lost, increasingly insecure in my relationship, fearful about slipping into old restrictive eating patterns, the dull ache of not knowing exactly where my life was headed. Some mornings, I woke up questioning the point of living at all.

But looking back, the first alarm came not from the brain I was working to the ends of the earth for this Ph.D., but rather my gut. The ache mentioned long ago was still always there, always a little bit dull and sat a little way below my ribs. My professor at the time was doing pioneering work on the effect of probiotics on psychosis and specifically schizophrenia, so in the spirit of scientific curiosity, I naturally started experimenting with probiotics. Each week contributed to a more or less consistent habit-stacking method. This was combined with rejigging my lifestyle in ways I never had, which, looking back, was funny because I thought I knew plenty, armed with my master's in Human Nutrition. Clearly, knowing how the body worked and how food worked was simply not enough to sustain behavioural habits that would last. As my gut health started to change, some signals became a lot clearer. The same ache grew clearer and started to burn. Long before I could articulate what was wrong, an intuition deep inside my gut continued to fire, as if my body was trying to scream a truth at me, but I refused to hear. It was during this time, after a couple of therapy sessions and arguments with my then fiancé, that I decided to make a series of bold choices. I broke off my engagement, turned down plans to move to

Germany, and started focussing on my health and reconnecting to the people who mattered the most to me.

Meanwhile, in my academic life, I was researching major depressive disorder (or MDD, better known simply as depression), focussing on biological markers and the physiological basis of low mood, that thing I felt like I had to push through every day. So ironically my own work hit closer to home than I cared to admit.

As I explored these questions and began repairing my relationship with food and my body, a pivotal moment arrived which, looking back, feels somewhat serendipitous. I met Nick Andreou, a fellow researcher and entrepreneur whose work originated with his own experience with irritable bowel syndrome (IBS)—a condition that affected not only him, but so many people we both knew. Our conversations often took place during long morning walks or late-night discussions, and over time, we started looking beyond digestion. We saw how deeply the gut and the brain were connected, and how this link shaped not only physical health but also emotional well-being. Nick would tell me how his anxiety worsened when his IBS flared. I recognised myself in that pattern. My own emotional chaos and gut discomfort often danced together in ways that felt impossible to untangle.

This partnership led me deeper into the science of the gut microbiome and how it influences mood, metabolism, hormones, and even body image. That exploration is what ultimately gave rise to this book, my career choices, and later, my personal brand.

This short book isn't just a reflection of my personal journey. Rather, it is a plan in action. It is an invitation to think about gut health not as a niche concern, but the foundation of how you think and inhabit and feel in your body each day. We begin by exploring the true reasons behind gut health: what you are actually experiencing when something feels off in your digestion, mood, or energy. From there, we explore how the gut, metabolism, and nervous system shape mood, emotion, and behaviour, often long before these patterns reach conscious awareness. You will see how inflammation, brain fog, and burnout often stem from gut–brain or gut–mood feedback loops that most health conversations fail to address.

The remainder of the book becomes a discussion on metabolic switches, things that often block people from feeling and looking as they'd like. Clarity of skin, better periods, smoother digestion, weight stability—these are what we aim to achieve and understand ways in which daily selection and biology interrelate. The aim is simple but not superficial: to help you build a daily routine that fits your body and your life, not one that only looks good on paper. When regulation comes first, clarity follows in digestion, in your mood, and in how you carry yourself in this world.

A New Way to See Yourself

This book is about a deeper, more universal truth: why so many people, especially high-functioning women here, feel stuck in cycles of bloating, brain fog, burnout, and body frustration—even when they're doing all the 'right' things. In *Gutsy,* I present a simple but powerful truth:

Your gut isn't just a digestive organ. It's the epicentre of your physical, mental, and emotional well-being. Your gut even shapes your mindset, and this changes the way you see yourself and what you believe you're capable of. This booklet weaves science with real, practical tools to help you align your gut, brain, and metabolism, so you can feel clearer, lighter, more energised, and more *you*.

Though I don't go in depth into it in this book, much of this work sits within a broader framework I developed called Mind-Gut-Metabolism. It was created to help women optimise how they feel, look, and live without shame or self-punishment. Health and vanity *can* coexist; caring about how you look doesn't make you superficial. Striving to feel strong, empowered, and clear-headed is a form of self-respect.

When you optimise your gut and mind, you gain more than physical wellness. You harness the tools that let you unlock the confidence to evolve your identity from a place deep, deep inside, and discover a totally new way of seeing yourself.

What to Expect

Gutsy is written for women who have spent years trying to do the right things and still feel slightly out of step with their bodies, women who care about health and are curious, who enjoy food and pleasure and beauty, and who are no longer satisfied with vague advice or one-size-fits-all solutions. It is for those who want to understand what is happening beneath the surface, rather than putting fragile Band-Aids over symptoms with supplements or rules.

Much of what is presented here is science, but it is science translated into personal lived experience, thanks to years of trial, missteps, and attention. I don't want to put a specific focus on aesthetics in this book, but when the body begins to work with you rather than against you, changes do follow. Appetite steadies, energy becomes more reliable, movement plans actually start to work and give you the body you never thought you could achieve, because the internal state finally has the conditions it needs to trust you. What follows is not a plan to follow rigidly, but a way of orienting yourself and a way of paying attention so that regulation and pleasure live hand in hand comfortably in your life.

Gutsy is structured in three tiers:

Tier 1: Reset—This first tier (chapters 1 and 2) is all about simplifying and observing your body while simultaneously building awareness. It helps you reconnect with your gut's early signals and begin understanding how emotional patterns and gut symptoms feed off each other.

Tier 2: Rewire—In this tier (chapters 3 and 4), you start to invite change. You'll learn to tune your biology through metabolic strategies, nervous system regulation, and food patterns that match your real and specific needs.

Tier 3: Reinvent—The final tier (chapter 5) invites a deeper identity shift. With the foundations and tools in place to nourish your gut and thus optimise the rest of your body and mind, you're no longer just managing symptoms and choosing from temporary solutions like cutting out entire food groups or taking a generic, off-the-shelf probiotic or supplement. You're building a personalised routine that will optimise you, for you.

Up to this point, you have met the story (my story) behind the science. You have a glimpse of how an organ most people only associate with bloat and digestion quietly shapes mood, identity, and the way a life unfolds. None of this is abstract for me; it has been written through very physical sensations in my own body: fog, heaviness, the ache in my chest after a meal, and then a sudden lightness when something finally clicked.

The next step is not to rush into fixing anything, but to learn how to listen.

Before any protocol, meal plan, or supplement, there is a quieter, more foundational task: noticing what has already been happening inside you. The subtle shifts before the crash, the way your stomach knots before a difficult conversation, the way your energy dips after certain foods or certain weeks—these are not random quirks, but information.

Tier 1 exists for that reason. Here, the gut stops being an abstract concept and becomes something you can feel and understand in real time and your real life. This is where you begin to trace very human experiences like burnout, reactivity, and that low-grade weariness back to their biological roots.

Think of this tier as orienting yourself in a room you have lived in for years with the lights off. The furniture has always been there, and the walls have not moved. You are simply learning where everything is and how it all connects, so you can orientate yourself much better.

Once that picture is clearer, the tools in the later tiers make far more sense—but first, the *reset*.

TIER 1
THE RESET

TUNING INTO YOUR BODY'S WHISPERS

And here we come to the starting point. This tier will make you hyper-aware of just how all-encompassing this one organ is and allow you to realise how our guts directly influence our energy levels, self-confidence, and the clarity of our thinking. It invites you to stop for a moment and rethink what has been fed to you about the human body. Of course, there is a heavier educational component here, without which it will be difficult to grasp the beautiful and complex nature of the body's internal state, most notably the gut. And this step really does matter, because understanding and education, the foundation of deeper knowledge, is the first step to change. Once you are able to grasp just how your gut communicates with your brain, regulates inflammation, and shapes your emotional landscape, you start to recognise and thus want to change patterns in your own life which you previously would have dismissed as random or 'just how I am'. You will learn about the gut–brain axis, how gut-based chemical messengers matter, and how the vagus nerve sends signals between your brain and your stomach. You get to understand how inflammation, stress, diet, and nostalgic emotional patterns interfere with your inner balance. So, *reset* is not about restriction or discipline but awareness of how to trace the burnout, emotional reactivity, or constant weariness back to their biological root. This new link between your gut and mental state opens a different kind of gateway. That is your reset, an awareness shift that lets you regain control of yourself, on which you can then start to build.

Key words:

Vagus nerve: a large nerve that connects the gut to the brain

Gut-derived neurotransmitters: chemicals made by the cells in the gut that can send messages to your brain and other parts of your body. Examples of these are also referred to as hormones, such as serotonin and dopamine.

Microbiome: the entire collection of microorganisms, such as bacteria, fungi, and viruses

Neuron: little messenger cell in your brain and nervous system

Molecule: small cluster of atoms bound together, making up everything from the water you drink to the hormones and chemicals your body uses and needs to function well

CHAPTER 1
UNDERSTANDING THE GUT-BRAIN AXIS: THE FOUNDATION OF IT ALL

In this chapter, I introduce the foundational idea that the gut is not just a digestive organ, but the command centre that underpins how we think, feel, and function. In fact, your mind, mood, and metabolism are all dynamically influenced by your gut and the microbial ecosystem, or trillions of bacteria, living within it. [3] Bear with me as I get through the necessary science terminology too.

The gut being related to anything other than digestion would be akin to science fiction just a couple of decades ago, so this is a pretty paradigm-shifting concept. When I first read and understood its all-encompassing nature, something clicked. I literally put my hand on my stomach, realising that all these gut-derived signals were not random quirks, but real shots of communication that most of the time I chose to ignore. In this book, we begin by exploring what the mind–gut–metabolism axis is, how the gut communicates with the brain and body, and why this bidirectional network may be the most overlooked key to lasting physical and emotional resilience. Remember that analogy above on moving about in a house where the lights aren't on? Think of this chapter as turning those lights on after years of trying to guess where the furniture sits. This chapter sets the tone for everything that follows, and acts as a launch pad for reclaiming your mood, energy, and metabolic balance from the inside out, quite literally.

One of the most profound reset moments I've had was a radical thought that changed everything for me, and one that may as well be one of the main theses of the book: *your gut helps govern your conscious experience.*

Consciousness, in rudimentary terms, is the ability to experience thoughts and emotions, the awareness that you exist. One of my biggest reset moments was recognising that the gut helps govern conscious experience. Consciousness, in its simplest form, is awareness: the ability to register thoughts and emotions, to know that you exist, to feel that life is happening through you rather than just around you. I still remember mornings when that awareness felt muted, as if someone had quietly dimmed the lights in my mind. It was only when my gut began to heal that those lights brightened, revealing what had been there all along.

Living consciously means going one step further. It's taking proactive control over how we interpret those thoughts and feelings. It's moving from reaction to regulation, and then action. And the gut, through its impact on mood, motivation, and metabolism, is central to that shift.

And this is not just something pulled from the annals of my diary, or those of the many women who journal through their moody monthly cycles. The science reflects it too. Mood disorders like depression and anxiety are frequently comorbid with gut-related conditions like IBS, which I earlier mentioned heavily afflicted Nick, affecting up to 10–20 per cent of the population. I've watched friends cancel first or second dates because their guts wouldn't play along. Not to say that their guts are wrong, but if your gut is inflamed and unregulated, the signalling happening between your gut and your brain becomes a lot harder to interpret. The famous phrases 'listen to your gut' or 'go with your gut' become less compass and more static radio, muffled signalling making your emotional reactions more outsized than they should be, or even totally unpredictable. At this point, you can really start to see how the gut influences such a core part of life, but we can get more into the science of that later on. Essentially, the gut-brain connection means that issues like IBS and mood swings often occur together, doubling the symptoms you experience. IBS is sadly often worse in areas with high socioeconomic stress, where inboxes don't sleep and sky-high levels of cortisol, our primary stress hormone, are the norm. Gut inflammation, which leads to microbial imbalance (an imbalance in our bacterial makeup), and impaired metabolism, are often happening at the same time, affecting our mood and thus how we feel and function. When the gut flares, you may feel bloated and uncomfortable, and this bloatedness oftentimes coincides with a tug in your chest, which ripples up to your thoughts. This is clear proof that the highway runs both ways.

The Gut–Brain Axis Explained

When I was in seventh grade, we were asked to detail the journey of a cheese sandwich, where, via gaudy pictures, we tried to draw the pathway a cheese sandwich went through in our mouth, oesophagus, stomach, small intestine, and so on. We learn about the different enzymes (proteins) that break down the different components of food—amylase in the mouth to digest some of the carbohydrate, pepsin or lipase for the cheese, etc. We were taught that the gut digests the food we eat and excretes the undigested stuff. It was a simple explanation without any further elaboration on its function. It is nothing more than a digestive engine, tirelessly working to break down food and absorb nutrients. That was the story my textbooks told, until fresh research flipped the page.

In recent years, science has revealed something much more profound: your gut can communicate directly with your brain, shaping your mood, emotions, and thus also the way you perceive the world around you, alongside metabolism-related parameters such as how efficiently you utilise food, how efficiently you can lose weight, and how well you can handle sugar. This connection is what we call the gut–brain axis. The first paper I read on it was perhaps my first reset moment—that jolt of clarity where something just clicks, and you can't unsee it from that point onwards. You really can think of this communication pathway like a control centre (the brain) within a bustling city (our guts) like London or Singapore: the control centre (brain) picks up information via messengers about the city (gut) and its citizens—criminal activity, pollution, birth and death rates, overall economic wellness. No wonder a dodgy snack can spark a full-scale riot inside me. The control centre integrates this information and makes decisions on what to do. Similarly, the control centre produces an output causing changes within the city to balance things out and optimise its health, be it more stringent criminal control, investing in infrastructure, or boosting the tourism sector.

A little bit of background on our guts, to reinforce just how all-encompassing this gut-brain axis really is: To give you an idea of the order of magnitude, we have about ten trillion human cells—think red blood cells, white blood cells, muscle cells, neurons (our brain cells)—and around 100 trillion microbial, or bacterial, cells. If that doesn't impress you, maybe the gene count will. The human genome contains around 20,000 genes, enough to encode all the necessary functions for life and death. Microbial genes contribute between two and twenty million. Their overwhelming numbers and genetic range show that my body is a shared ecosystem rather than a purely human one. That dwarfs our count a hundredfold, and we haven't even discovered all the bits of our microbiome yet. To make it clear, I am referring to our total microbiome, including the microbiome of our gut (the gut microbiome), our mouths (oral microbiome), and other places such as our genitals (the vaginal microbiome).

The gut–brain axis is this extraordinary communication highway that I'm convinced would be considered some kind of science fiction if not for all the research that's exploded in the past few years. To make it easier to understand (aside from the crude control centre example mentioned earlier), the way this communication occurs can be divided into three parts: neuronal, immune, and chemical. While these pathway constantly overlap, separating them helps clarify how signalling occurs, I explain this using serotonin, our feel-good hormone, as an example.

Immune: When Your Gut Goes on Alert

Any time I pick up a cold, my stomach throws a tantrum—proof the gut reads the news before my nose does. This is what's going on: the bugs in your gut, and the cells that line it, or *enterocytes*, chat non-stop with your immune system. That back-and-forth decides how well you fend off germs and bounce back from them.

Gut microbes and these enterocytes lining our gut produce a whole range of molecules, signalling molecules, most commonly called cytokines, which act as the body's messengers that regulate inflammation. When the gut is inflamed, these inflammatory signals can travel through the bloodstream and influence the brain's immune response. This then leads to neuroinflammation, or inflammation of the brain, which has been strongly linked to mood disorders such as depression and anxiety.

Another very clear and intuitive example of the gut–immune connection is that of skin issues. Conditions such as acne, rosacea, and eczema are tightly linked to the gut–immune pathway, otherwise known as the gut–skin axis, and inflammation of the gut, which continues talking to our immune system, manifests in a variety of these skin issues. When the immune system is activated, inflammatory signals are sent to the skin, including your oil glands and follicles, triggering skin-related symptoms.

I had awful childhood eczema growing up, which I talked about in a casual panel recently alongside a naturopath, TCM (traditional Chinese medicine) practitioner, and somatic practitioner. It was eye-opening just how much overlap we found across our disciplines. A shared belief and consensus of the panel was that inflammation, which is caused by over-activation of the immune system, mirrors what TCM describes as 'dampness', or stagnation of fluids in the body. I described my own experience and journey with eczema: I had to be wrapped up like a mummy by my mum most nights to prevent myself from constantly scratching away at weeping, itchy sores. Looking back, though, despite feeling slightly embarrassed by my skin, I was a child without a care in the world—all I wanted to do was read books and play with my neighbours outdoors. Playing a lot in soils, fields, and playgrounds—exposed to a broad mix of microbes in the environment—eliminated this eczema and my allergies (sashimi, cats, and dogs, to name just a few) almost overnight.

It was quite beautiful to observe that overlap between an abstract concept of heaviness and dampness, rooted in a very real and scientific connection between our guts and immune systems. I left the health panel feeling more connected to other healing traditions, with a deeper appreciation for our gut as seated at the centre of disciplines we would otherwise consider unrelated. As we discover more and more about this magical organ, the more undeniable it becomes that digestion is the tip of the iceberg, only hinting at the many ways our gut impacts the way we feel and function.

Heading briefly back to the topic of mood, which is a central theme here, since mood affects everything from the way we look at ourselves and our inner potential, to the way we set up habits to inspire behavioural change. When my gut feels like a bonfire, my mood often tanks right alongside it. Learning this was a reset moment: realising that mood swings and immune dips weren't just in my head, but rather quite literally from my gut.

From an immune pathway point of view, serotonin—a chemical messenger involved in regulating mood, appetite, sleep, and motivation, of which up to 90 per cent is made in our guts, and one which we revisit in the next section—also plays a role in inflammation levels, with downstream effects on brain function and mood.

So how do pro-inflammatory cytokines, these tiny messenger proteins, influence mood? These cytokines disrupt the metabolism of serotonin by degrading the bigger chemical that it is broken down from, called tryptophan (which you will find in things like chicken and eggs). That crash you feel after a junk-food binge? Part of it is this chemical traffic jam. Left unchecked, long-term inflammation can stiffen the brain's wiring, making new habits or fresh mindsets feel like pushing through wet cement. Realising this was yet another reset moment for me—understanding that what I ate wasn't just affecting my waistline, but my ability to think clearly, feel motivated, and change.

Chemical: Quiet Currents Beyond Mood and Metabolism

The second pathway is the chemical one, and a quietly powerful one at that. It involves dozens of chemical messengers—including serotonin mentioned in the previous section—used to regulate mood, cognitive sharpness, and even appetite. Dopamine is an important chemical messenger here alongside serotonin, one which you have probably heard of before, the hormone that gives you motivation to do things like get out of bed early and finish writing parts of a book with content that resonates close to your heart. These neurotransmitters are produced by our gut bacteria, so in a way, these microbes are manufacturing the very building blocks of what helps us feel like ourselves.

Therefore, when the balance of gut bacteria is disturbed, the production of these neurotransmitters is also altered, contributing to emotional imbalances or a sense of feeling off. As mentioned, the gut produces up to 90 per cent of our body's happy hormone serotonin, which already tells you how central the gut is to our emotional lives.

Another essential piece of the chemical pathway is dietary fibre. When we eat this fibre, it is broken down by enzymes produced by our gut bacteria, then fermented to short-chain fatty acids, or SCFAs. Now these compounds are the unsung heroes of our guts. SCFAs help reduce the production of pro-inflammatory cytokines and strengthen our gut linings, preventing the 'leaky gut' phenomenon, where unwanted toxic molecules slip into our circulation and cause systemic chaos. Understanding the power of SCFAs was another reset moment for me. They are like little janitors sweeping through the corridors of my gut, extinguishing all the unwanted fires of inflammation.

This communication is what makes the gut–brain axis so endlessly fascinating. It's not just about digesting those cheese sandwiches anymore; rather, it's about how your gut health shapes your mental state, the way you feel, move, and exist in this world. Whether you're aware of it or not, your gut is constantly talking to your brain, influencing how you react to the world around you, alongside your motivations, cravings, and the overall tone and voice of your inner life. Realising this might be a reset moment for you, just as it was for me: you stop seeing your gut as just a digestion machine and start recognising it as a powerful driver of your mood, focus, and daily experience.

Neural: How the Body Feels Safe

We end with the neural, not because it matters more than the others, but because it is the most immediate. This is the more direct line of communication between the gut and the brain. That does not mean the immune and chemical pathways do not affect the brain as well—they do. But the neural pathway is different in that it allows the gut to speak to the brain in real time.

At the centre of this relationship is the vagus nerve—a long, winding, surprisingly elegant nerve that acts as the main highway between the gut and the brain. Think of the vagus nerve as the hotline my stomach uses when it needs to send an urgent memo upstairs. It's the longest of all our cranial nerves, extending past the brainstem and innervates other major organs in our body, including the heart and liver.

The end of the nerve has these receptors that pick up sensory information in our guts. They detect changes in the microbial environment, presence of food, hormone levels, and relay this information up to the brain to be picked up and processed by parts such as the nucleus tractus solitarius (NTS) in the brainstem. Through this nerve, the gut sends signals to the brain, telling it how things are going down below. It's a two-way street too: when your brain is stressed, it affects your gut via signalling to the vagus nerve; when your gut is inflamed or imbalanced, it sends distress signals back to the brain, contributing to emotional disturbances.

Robust clinical data back this direct pathway. Implanted vagus nerve stimulation (VNS) has been shown to provide sustained relief in treatment-resistant depression. [4] They did the same thing in a totally non-invasive way too—sending mild pulses through the ear to activate the vagus nerve, which feels nothing more than a light tingling, was shown to ease pain and improve motility in irritable bowel syndrome, and lowers inflammatory activity in paediatric inflammatory bowel disease, positioning it as a promising gut–brain therapy. [5, 6] More than anything, you can see just how central this one large nerve is to mediating communication between the gut and brain.

The overlap between the aforementioned chemical and neuronal pathways here becomes especially clear through the vagus nerve. This nerve, which we were introduced to in the first section, senses how much serotonin (or other neurotransmitters, and the examples here may also be labelled as hormones) is present, relays this information upwards towards the brain, and helps the brain calibrate mood and stress responses accordingly. It was the core mediator in the neural section for a reason; these pathways don't operate in silos. Rather, they braid together. Serotonin itself is produced in specialised cells embedded in our guts, and levels vary depending on stimuli in the gut, such as how much or the types of food present, or which microbes are thriving. It's important to note here that serotonin does not literally travel to the brain and enter it. Our brains are protected by a barrier called the blood–brain barrier, which filters what can enter and exit our brains. Although serotonin does not directly pass through the blood-brain barrier, it does indirectly influence our mood; high levels of serotonin in the gut can promote feelings of well-being by activating sensory neurons in the brain via the vagus nerve—indirect but powerful.

The Gut as a Second Brain

Scientists now call the gut our second brain, and honestly, it often feels like the first one to complain. What may have once been dismissed as pseudoscience just a decade or two ago is now grounded in cutting-edge research. For many of us, learning this can be a *reset* moment: the instant you realise your gut isn't just along for the ride, but actively shaping how you think, feel, and function every day. This sudden surge of books, courses, and studies on the gut–brain axis is not just a passing trend—it's a direct reflection of how integral our guts are to both our mental and physical health.

As you can see from the description of those three core communication pathways (chemical, immune, and neuronal), there exists not just this simple relationship between gut health and brain health. Gut microbes are, in fact, direct mediators of your mood, susceptibility to immune disorders, executive functioning, and even skin health, factors that do not result from mere lifestyle or circumstance. As projects such as the American Gut Health Project have discovered, individuals with clearly dysregulated gut microbiomes often experience heightened levels of anxiety, depression, and cognitive dysfunction, which is to say, yes, your jobs and friendships do play a large part in how you think, feel, and show up in your life, but your gut shapes these long before you may consciously register it.

At the end of each chapter, I'll ask a question to make you slow down a little, just enough to make you notice something that you may have rushed past if not for some degree of introspection.

End-of-Chapter Reset: *Now that you have seen how dynamically the gut and brain communicate, what is one insight from this chapter—scientific or personal—that changes how you see your body's inner world?*

The next time you sit down to eat a meal, pause for just a minute before the first bite. See if you can sense the slightest signal, be it hunger, irritation, tension, excitement, or even a swirl of feelings you can't quite name. It doesn't have to be profound in any way, just honest, because those signals are exactly what we will work with next: the emotional messages your gut has been delivering all day long, long before you learned how to properly listen to them.

CHAPTER 2

THE GUT AS AN EMOTIONAL FILTER (THE RESET TIER)

Every strong feeling I've ever had, be it joy, dread, or first-date jitters, has announced itself as this mild ache just below my ribs. Can you recall a recent time when you were stressed? Did you have a sudden stomach-ache? Or perhaps one might say one has butterflies in one's stomach when nervous about an impending exam or first date. These are not mere coincidences; your gut and feelings are more intimately linked than you ever thought.

My early interest in psychiatry was born out of this very tension. I was fascinated by behaviour, mood, motivation, and how poorly we seemed to explain them through willpower or cognition alone. That curiosity eventually carried me into my doctorate, but it unfolded alongside periods of my own psychological unrest, times when change felt strangely unreachable. I could eat well, exercise, do all the things I was supposed to do, and still find myself repeating patterns I understood but could not interrupt. Nutrition education alone could not account for this gap between knowledge and behaviour. There was clearly more at play.

This section exists to sit inside that space, to look at the gut not as a passive organ, but as an emotional filter that shapes how we experience the world, how we react under pressure, and how we form habits long before conscious thought enters the scene. Before optimisation, before intervention, there has to be a reset: an acute understanding of where emotion is first felt, how it is processed through the body, and why the body so often responds before the mind has words for what is happening.

Emotional dysregulation is one of the most common and quietly frustrating struggles people face today, particularly women. Our emotional highs and lows are not random, but deeply intertwined with hormonal shifts across the menstrual cycle. What often goes unnoticed is that these shifts are filtered through the gut, where hormones, microbes, and immune signals meet before their effects are felt in how we feel and show up in the world. Anxiety, low mood, irritability, and burnout do not live in isolation from the body. They emerge from the same gut-driven pathways that regulate metabolism, energy availability, and resilience. When the gut is strained, the fallout is rarely contained.

The Magical Intelligence of Females

I hear the words 'gut feeling' a lot more from women than men. Many men are healthily in tune with their emotions, which overlap with our gut function, but I would still argue that such a phrase originates a lot more from my female peers. 'I have a good feeling about him' or 'I know a good thing when I feel it in my gut' is something I hear way more often in my female friends and family than in males, and it is hardly accidental.

From a young age, girls learn to read nuance. We make eye contact with each other across the dinner table when an uncle makes a tactless remark; we feel the little shifts in tone and expression in our friends' text messages. We intuit silence and anticipate conflict before it even arrives, because we are literally biologically wired for it. Since the gut is so densely wired with sensory neurons, coupled with the fact that women tend to have more finely tuned interoception (the ability to sense internal bodily signals), the gut often becomes our first emotional barometer. This is also why it is common in a relationship to assume our male partner (assuming you are in a heterosexual relationship) should be able to read our minds and figure out how we feel and think long before it is made obvious. And should we be surprised about this? Up until toddlerhood, a woman has to interpret the nonverbal cues of her child. In fact, many mothers would say they sensed their baby's rhythms and needs long before birth: the shifts in movement and kicks in her belly, knowing when something feels off despite normal scans. Birth only sharpens this intuition radar. I daresay such an intuition does not lie languid in our psyche, rather moves through us and pulses in our blood as this active expression of our gut health. I met one of my very best friends in London through a shared interest in lash extensions before a lecture at Imperial College during my master's in Human Nutrition. Something about the way she approached the questions, her energy and laugh, the way she moved through class, all made me drawn to her, and years later, I found out that she felt the same way about me. Both of us had a gut instinct that this friendship was one to last, but it was maddeningly sad when I decided to move back to Singapore to pursue an exciting new job. Despite being separated by thousands of miles and plenty of hours, it was difficult to untangle the friendship, and many a time, I found myself in a situation where I would send her a text message, and she would reply, saying, 'That's so strange, I was just thinking about reaching out to you.' It was as if our guts were held together by some unseen thread of connection, one that distance couldn't fray. It's been almost a decade since we have been tied together in both heart and mind.

I could very well say the same about my romantic relationships, though this is a tricky one because ideally, you would need to have worked through some past trauma and understand what you really need before diving into one. In the case of trauma, what feels like intuition can simply be the body reacting to familiar emotional patterns rather than present-moment truth. Trauma-driven signals tend to be emotionally driven, with feelings of passion and excitement combined with that of anxiety or stress, but a gut feeling is more knowing, steady realisation, rather than a dramatic surge of passion.

Since my late teens, I had experience with transactional relationships, a manifestation of a decoupling between my body, specifically my gut signals, and mind. Many late nights saw me opening doors and entertaining rich, older men in exchange for massive financial gain. What emerged was a sort of adaptive dissociation: a learned ability to override discomfort in pursuit of a desired outcome. For a short time, I deluded myself into thinking that I was happy to live like this, but soon I came to a dangerous realisation that the cognitive override, emotional compartmentalisation, and suppression of interoceptive discomfort all swirled together with regular bottles of alcohol and extra cash served as fuel for my binge–restrict episodes. I was highly functional and outwardly in control, yet increasingly disconnected from the internal cues that might have guided me towards greater alignment. My poor gut did not disappear; it was simply trained not to interrupt. It was only after I began restoring that internal dialogue through nervous system regulation, nutritional, and consequently gut repair, and a more honest relationship with my body that something shifted. Around this time, I met my partner. And from the very beginning, the signal felt different.

I downloaded a dating app when I first moved back home to Singapore, with zero expectations or flights of fancy. Having just emerged from a rather toxic relationship back in London, I was ready for a clean slate and hoped to find a meaningful connection with someone I was attracted to. I had a whole list of embarrassingly specific things I expected from a romantic partner: kindness, passion, affection, a certain family background, preferably lived alone, financial stability, ambition, handsomeness, God-honouring—it goes on.

When I came across my now fiancé's profile, there was an undeniably strong gut feeling, this quiet sense of knowing and understanding, tugging away again, eerily similar to the tug I felt when I realised I had to break off my engagement for the long-term good of my romantic career, years ago. At first, I was sure I was swayed by his tall, handsome profile and nothing more, but the dates that ensued proved this strong gut instinct to be right all along, and it was made evident from his end as well. After a year of dating, we were engaged.

Some call it the right energy or the right vibe, but these arise from the gut being both a predictive organ and an emotional processor. It is constantly integrating sensory cues, past experience, and present physiology into this deeper knowing that precedes rational thought. This gut instinct saw us through minor trembles in the relationship too, especially in the beginning when it felt easier to walk away than stay. It was clear that these trembles originated from me having to confront my own fears and weaknesses, and his own gut instinct made him work through his own.

Sex and libido is an important part of my life, one that shapes romantic relationships inevitably. While working at Amili, Southeast Asia's top precision microbiome company, I discovered emerging research on the oestrobolome, or clusters of bacteria influencing female hormones, namely oestrogen and progesterone, and thus libido. Of course, my partner was attractive, but the pull I felt towards him ran deeper than the kind of surface-level, animalistic attraction one might experience towards any good-looking stranger in a crowded room or dimly lit bar. There was a sense of recognition, of calm certainty, as though my body had reached a decision long before my mind had finished analysing the facts.

It is so clear that we have this subtle intelligence that we carry around in our tissues—the fog in your thoughts, the tug in your chest, that female intuition, which I may argue is in itself a magical intelligence that the other sex will never know or quite understand, but will attempt to anyway, as I can attest to as a straight female who has been in many different heterosexual relationships.

Let this be a *reset* point for you: a moment to rethink what your body has been trying to communicate with you. By tuning into these signals, you initiate an awareness process that, in return, initiates a paradigm shift from blindly responding to emotions towards recognising what they are about. This awareness sets off another reset of how you go about managing your stress, relating to others, and genuinely caring for your emotional well-being.

Gut Inflammation: The Silent Mood Saboteur

Let's talk about inflammation for a bit, something often blamed for all kinds of health problems, but rarely connected to mental health. Chronic gut inflammation, often caused by poor diet, stress, or an imbalance in gut bacteria, doesn't just stay in the gut—it can travel up to the brain, causing neuroinflammation, [7] a term we briefly mentioned above, when making its link to mental health issues and mood disorders like depression and anxiety, which happen to be comorbid with both gut and metabolic issues, such as eating disorders and IBS. These are clear examples of how gut inflammation can spill over into emotional health.

I never paid much attention to how I felt growing up; it was seen as a distraction from achieving 'the important things' like getting good grades and getting into a good university. Feelings were just feelings—natural, fleeting, and in the eyes of typical Asian parents, not something really worth indulging. I quickly learned how to stop crying on command, halt pursuing arguments, or shut them down midway with boyfriends if I felt that wince of annoyance or frustration (because shouldn't they know how to act without me telling them?), to keep it all in until it boiled over into eruptions of anger, which seemingly came from nowhere. Emotional control became my currency for survival, to stay in control no matter what life hurled my way. Years later, my therapist diagnosed me with alexithymia, which was a pretty fitting name for the inability to identify and differentiate between emotions.

During the transition from breaking off my engagement and rebuilding a life that actually felt like mine, I was doing a lot of both emotional and biological experimentation. It was a worthy effort trying to piece together the most holistic lifestyle I could manage: targeted probiotics, supplements that made sense instead of a bunch of random wellness noise on social media, balanced meals, long daily walks, and therapy sessions that felt uncomfortable but were at least honest. We will unpack the methods and plans of action later on, but here's the important part: As my gut changed and as it started to regulate itself, the clearer my emotional dashboard became.

That clarity was yet another reset moment, and the shift was impossible to ignore. The more my gut calmed down, the more clearly I could think, feel, and decide. This was perhaps the first time I truly understood that emotional healing wasn't just about mindset, but about my microbiota too.

This realisation eventually became the backbone of the framework for my philosophy, which introduces a new way for women to understand, with both science and self-trust, how to take control of their bodies and brains to look and feel their absolute best.

Your Inner Compass: Mood Through the Gut

Mood serves as the lens through which we experience the world, colouring every thought, action, and interaction. It shapes our interpretations of life's events and dictates our responses to them. This was the theme of a friend's podcast I was on a few years ago, discussing the connection between gut health and mood. This podcast was interesting because it was specifically in the context of men's mental health, and at the time, he was working at a men's mental health start-up. Most men are conditioned not to speak about the way they feel, because of both societal pressure and past trauma, and not to say that they lack the gut intuition females have; rather, it appears easier for these societal pressures and traumas to mute this sharper signal between healthy emotional processing and gut health.

Through Amili, I pitched the science-backed products to my doctors and, in doing so, became friends with them and shared many a story. I recall a particular lunch conversation with a prominent psychiatrist, Dr Haridas from Novena Medical Centre. We touched on many topics but landed on the Bayesian brain theory, which I credit Dr Karl Friston for. This theory posits that the brain is not merely just this filter that processes incoming information—rather, in quite a computer-sciencey way, constantly updates our belief systems and ways of looking at the world by combining past experience with new sensory information to make predictions. The idea is that (what you expect) - (what actually happens) = (prediction error), and the brain works hard to minimise prediction error to learn, act, and think more efficiently. In the framework of psychiatry and his patients, which we touched on inevitably since I worked in the psychiatric department for graduate school, the hypothesis is that depressed patients have very negative and rigid priors, or the first part of that equation, on what they expect. Examples of these rigid and negative priors may be repetitive thinking around such lines as 'Nothing will get better', 'No one cares about me', and 'I'm so stupid and incapable'. This lines up with the core symptoms of depression or low mood: the inability or reduced ability to feel pleasure or interest in activities this person previously may have loved doing, a pervasive sense of flatness or dullness, and difficulty concentrating or making decisions. This negative characteristic of low mood makes the prediction error get wider over time, because even if a positive reality occurs, their internal model cannot integrate this. Over time, as the prediction error gets wider and larger, everything from this person's emotional to cognitive state starts to become misaligned.

I could not help but link this same model to the experience of a strong gut feeling or intuition. If the Bayesian brain theory suggests that our brain is constantly comparing prior internal models about the world with incoming sensory data, then our gut feeling can be seen as the moment when these predictions line up nicely, so the body generates a preverbal sense of *something feels very right, or very wrong*.

In other words, our gut instinct may be the felt experience of low prediction error, not high, as in the case of depression or even autism, where there is a failure to understand the external world well and, in a way, their body's internal states too. Our gut feeling, then, is not something mystical at all, but rather the body's embodied recognition that the world in front of us is coherent with what we've learned, lived, and sensed before. It's important to note the distinction between mood and emotion—mood is a longer, more chronic state that arises from this prediction error, and emotion is the short-term feeling that stems from the local mood. Anger, for example, may arise from a mood state coloured by chronic threat perception, where the brain has learned that the world is dangerous and unsafe. Mood is a bit like the ocean and its undercurrent; emotion is like the waves that flit and dance at the ocean's surface. When the gut is dysregulated, the ocean and undercurrent start to destabilise, and emotions follow suit in turbulence.

Furthermore, as research increasingly demonstrates, mood is not merely a by-product of circumstance—it is profoundly influenced by stress. [9, 10] Stress doesn't simply unsettle us temporarily; it rewires how we think, feel, and see the world. A bit of stress is beneficial, even exciting. I enjoy cooking for small groups and planning a getaway. These are all examples of relatively exciting, short-term, positive stress events. Working as a scientific liaison at Amili, I derived genuine pleasure from following up with doctors and clinic managers to strike close and inform doctors how and why targeted probiotics and even faecal microbial transplants (transplanting a healthy gut microbiome into a diseased one) strengthen the gut–brain axis and thus overall internal health and well-being. However, chronic stress—the kind that lingers for extended periods with unresolved outcomes—is a different matter. It clouds our mood, leaving it murky and distorted. Its effects ripple through both physical health (notably the gut, of course) and mental well-being.

We're now approaching a very important part of the book, one that cements one of the strongest knowledge pillars in this realm: Stress wreaks havoc on your physical health (including your gut, of course) and mental health (mood). Stress doesn't just influence your day-to-day emotions; it has the power to disrupt your overall well-being. Here, we explore how understanding and managing stress is key to preserving not only your mood but your entire health ecosystem.

How Do I Know If My Mood and Emotions Are Well-Regulated?

I start with one clue: do I wake up curious or cranky? That gut check tells me more than any wearable can. The densely interconnected regions of the brain that regulate our mood and emotions highlight the robust scientific foundation of mood regulation. The brain regions which control our mood, or how we feel, are: the prefrontal cortex, [11] amygdala (part of what is called the limbic system), [12] nucleus accumbens (NAc, our main reward system), [13] and anterior cingulate cortex (ACC). [14] Together, these regions act to influence our emotional reactions to our everyday realities and life's challenges. If I were to elaborate slightly further, optimised mental health may be defined as a feeling of subjective well-being, perceived self-efficacy, autonomy, competence, generational dependency, and self-actualisation of one's own intellectual and emotional capacity. That may all sound like a different language, so imagine this instead: You have a friend named Emma, who is a writer and editor. She wakes up every morning with a sense of purpose, feeling content and energised (subjective well-being). Emma takes pride in her work, believing in her ability to meet challenges and solve problems that come her way (perceived self-efficacy). She has the freedom to choose the projects she works on and how she manages her time (autonomy), which makes her feel in control of her life.

Emma continually hones her writing skills, improving with each project (competence). She may make mistakes, of course, and receives negative feedback every now and then. She wakes up some mornings not feeling tip-top and gets into arguments with her parents and partner, which make her question things in her life, such as her finances and relationships, especially during that time of the month! However, she is resilient enough to overcome bouts of failure or feeling 'lesser than', because she understands that her value lies well beyond her occasional slip-ups. She values the wisdom shared by her parents and mentors, knowing that their experiences contribute to her own understanding of life (otherwise called generational dependency). Emma also sets time aside each week for activities that push her creative and intellectual boundaries, such as taking courses in philosophy, trying a pottery class she's had on her creative to-do list for a while, or engaging in deep conversations with friends who uplift and inspire her. This, of course, helps her fulfill her potential and grow as an individual (self-actualisation of intellectual and emotional capacity). This is a whole and fulfilled person. We all want to be, in some measure, like Emma. We want to be able to realise our potential and work on our skills and surround ourselves with people who see this and encourage this in ourselves too.

Stress and the Gut: The HPA Axis

Your brain has a built-in fire alarm system. It's called the hypothalamic–pituitary–adrenal (HPA) axis, and it's how your body reacts when life throws stress your way. This system connects three key players: the hypothalamus (in the brain), the pituitary gland (also in the brain), and the adrenal glands (which sit on top of the kidneys and release stress hormones in response to stress). [15] Imagine this HPA axis as your own internal fire alarm system. When the hypothalamus senses danger, the alarm is pulled: the pituitary and adrenal glands respond by releasing stress hormones like cortisol. One of the key messengers here is a hormone called corticotropin-releasing hormone (CRH), which works alongside oxytocin, also known as the love hormone, as it is released during and after intimacy and makes you feel calm, connected, and overall really good. [16] Together, these work to regulate our stress response and influence mood by interacting with various brain chemicals. Our handy HPA axis is activated whether you're frozen in front of a bear in the forest or thinking of how to respond to yet another rude and unnecessary question from a family member at Christmas dinner.

How do we know this axis is even real? Well, when it's out of balance, we do see the consequences—dysregulation of this axis is often linked to mood disorders like anxiety and depression, showing just how deeply wired it is into our emotional health. [17] Hormonal pathways, involving oestrogen and progesterone, also interact with this HPA or stress axis, thus influencing how we respond to and manage stress. That is why fluctuating hormones, such as those implicated during that time of month, levels of oestrogen cause mood swings during that special time of the month.

But none of this starts in the brain alone. It starts in your gut, where our tiny microbes help make neurotransmitters: chemical messengers that affect everything from motivation to anxiety. Serotonin, the one we all hear about, is mostly made down there. As mentioned previously, up to 90 per cent of the body's serotonin is produced in the gut, not the brain. [18] So when your gut is out of balance, it can have a significant impact on how you feel emotionally.

I used to ignore this connection, brushing off my gut's complaints as unrelated to my mood. But once I understood how deeply connected they were, I couldn't unsee it. The body speaks—even when you're not listening.

Female Physiology Demands a Different Lens

A special mention again to all the women out there, who must navigate the hormonal shifts, metabolic fluctuations, and emotional tides that come with a cyclical body. We are truly silent warriors. I'm sure we women know all too well the effect hormones have on us, every single month! Our mood and emotional resilience fluctuate with our cycle—drops in the hormone progesterone during the luteal phase mean that we are more liable to grumpiness, and we tend to feel happier, more social and energised during our ovulation period, when oestrogen and testosterone levels peak. [19] I used to think cycle tracking was too fussy. Now I see it as basic self-respect. It helps me make sense of what I'm feeling and gives the people around me a little context too. Knowing how your hormones affect you day to day allows you to treat yourself with more kindness and to understand why you might feel less upbeat on certain days. It's a win for you and a win for those around you, who may be affected by your ups and downs. I'm not saying you should attribute every bout of low mood to your cycle, but it's at least something you can take into account.

Across different ages, many women experience similar hidden frustrations, whether it's pre-menstrual syndrome, perimenopause, weight gain, or the emotional weight of feeling invisible as we grow older. The challenges do shift with age: a 25-year-old might be battling burnout and career anxiety, while a 45-year-old may be confronting hormonal chaos, metabolic changes, and deeper questions of identity. I may not be 45 yet, but as other women may attest, even between the ages of 20 and 30, you do feel a difference in the way you look at yourself and how you feel internally and relate externally to the world around you. Your cycle, during the years that you still have it, also matters a lot. Painful periods, dull-looking brownish blood, heavy or too-light bleeding, and overly short or long cycles are all indicators of something not being very right, and a lot of this is caused by, yes, stress. [20] Perhaps this isn't so surprising since we as women tend to have a lot on our plates to juggle, especially between our mid-twenties until menopause. We are expected to find the right partner, maybe have kids, find a career that's meaningful to us, all while our hormones rage on and on, every single day of the month. It is simply something the male population cannot understand, so we must have the right tools to manage this ourselves and forge forward strong, but maintain a degree of softness to feel happy and content with our lives.

This is where tailored support matters. Through my philosophy, I help women understand their unique hormonal blueprint and the connection between their gut, brain, and metabolism so they can reclaim control, feel stronger in their bodies, and start living in sync with who they envision themselves to be. No more one-size-fits-all advice, just the right tools for the right phase, at the right time. I go into some diet and lifestyle advice in the practical application section of this book, down below.

Because when you work *with* your body, not against it, you don't just feel better. You evolve.

End-of-Chapter Reset: *Reflect on a recent emotional pattern. Can you identify a corresponding gut symptom or stressor that might be feeding it? What small observation will you make tomorrow?* You don't need a big revelation. Just one honest moment of connection between what you felt and what your body showed you.

You've started to listen differently to your gut, your moods, your signals. Now it's time to take that awareness and build new patterns that work *with* your biology, not against it. Let's move from Reset to Rewire.

TIER 2
THE REWIRE

EATING AND LIVING FOR INNER HARMONY

The key value of this tier is that it puts awareness into action. Rewire teaches you how to unlearn or recondition some of your emotional and physical responses through diet and lifestyle shifts in support of the gut–brain connection. If Reset revealed how and why your mood and metabolism have fallen off balance, Rewire is going to tell you how to manufacture that balance back again. This is what will allow you to update some of the emotional constructs and biochemical loops that no longer serve you, whether it is insatiable cravings for doughnuts every time you hit that time of the month or killer mood swings that are spilling over into the most important relationships around you.

The whole notion of diet is introduced here, but not in the superficial sense one might imagine. I actually resent the word *diet*, for it conjures up those old ghosts of restriction, too many rules, and the very confusion we are trying to eliminate. Here it is reframed as a holistic practice instead of a restrictive one, one inevitably intertwined with the workings of gut health but much more expansive than just ticking off a checklist of what to eat. You shall see that food is much more than just fuel; it is a highly potent feedback system. What you eat literally rewires your emotional tone and focus, along with your energy and cravings. Through simple and strategic shifts, you can consciously implement simple changes that result in a whole-body experience of feeling safe, calm, and emotionally resilient. Rewire puts into your hands the tools to respond consciously rather than react from old stress patterns.

Key Words

Metabolites: Tiny chemical by-products made during digestion and metabolic processes that send signals affecting your mood, energy, and inflammation levels.

Tryptophan: An amino acid found in foods like eggs, chicken, and tofu that the body converts into serotonin, a key mood-stabilising molecule.

Short-chain fatty acids (*SCFAs*): Substances produced when gut bacteria break down dietary fibre, which help regulate inflammation and mood.

Kynurenine pathway: An alternative breakdown route for tryptophan that's triggered by inflammation, which reduces serotonin levels and may worsen depression or anxiety.

Neuroinflammation: Inflammation of the brain caused by signals from the gut, linked to cognitive issues and mood disorders.

CHAPTER 3
EATING FOR EMOTION: THE DIET-MOOD CONNECTION

You've probably heard the saying 'You are what you eat'. But what if I told you that you can *feel* what you eat? We've talked about how your gut and brain are in constant conversation. The stress you feel, emotions, and mood all tie back to what's going on deep in your gut, thanks to the trillions of bacteria that live there and are actively and constantly conversing with the brain via the vagus nerve, chemical messages such as neurotransmitters and metabolites, and immune-mediated signalling. Now we finally get to the part that lands on your plate: food.

Food Is Information

Food isn't just fuel. It's data. Every bite you take delivers signals to your body: turn on this gene, suppress this inflammation, produce that neurotransmitter. [21] In a way, it's like receiving a text from someone you fancy—the same way that a single text can make you smile or spiral, food messages your cells and your microbes. These messages shape how you digest, how you feel, and even how you think. Think back to the last time you sat down to enjoy a well-balanced meal with a loved one. It almost feels like a reset button, doesn't it?

Certain fibres, for example, act as prebiotics, which are nourishment for beneficial bacteria. When those bacteria are well-fed, they reward you by producing short-chain fatty acids (SCFAs), the mood-boosting metabolites that support gut barrier health and reduce systemic inflammation. [22, 23] On the flip side, diets high in ultra-processed foods feed the wrong kind of bacteria, or worse, starve the good ones. This fundamentally shifts the gut ecosystem to dysbiosis, gut barrier dysfunction (also known as a *leaky gut*), and sets off biochemical cascades that influence emotional regulation, stress reactivity, and your cravings.

You may by now already know what a processed food is. They are foods chemically changed to preserve their shelf life, reduce manufacturing costs, and enhance flavour stability. Ultra-processed foods, sadly, are everywhere. They don't just include those doughnuts you crave that time of month, but literally any food that has non-natural ingredients in it, including artificial sweeteners, additives, preservatives, stabilisers, and emulsifiers. Flip the food package and scan the ingredient list—anything processed would have ingredients that your grandparents would not be able to identify. They dominate the aisles stacked along the middle of the grocery store and the small shelves flanking the cashiers, because don't we all need additional fun-sized M&Ms to top off our week's list? It's one thing to occasionally enjoy some chocolate; sometimes I'm impressed by the lengths gone to invent the chemical makeup of a simple store-bought condiment, even. Layers of millefeuille cosying up to softly whipped chocolate mousse and chocolate fudge in a birthday cake, sushi rolls stuffed with artificial crab made by your favourite hotel patisserie, for example, is food for the soul.

The issue isn't enjoying these processed foods occasionally; it's when they become the backbone of your diet and your management system behind it. When they dominate your breakfasts, lunches, snacks, and dinners, you distort the gut's signalling system, causing a more fragile mood and swinging cravings. It is particularly difficult to just have a small amount of processed food, since they are biochemically manufactured to light up specific parts of the brain with the golden ratio of sugar to fat to salt. These include the nucleus accumbens (NAc), which encodes cravings, making you inhale whole bags of chips without even realising it. There is the ventral segmental area (VTA), which projects dopamine to the NAc and your prefrontal cortex and makes you feel both fixated and uncontrolled around a food, highly implicated in binge-eating disorder. There is also the amygdala, which assigns emotional importance to your Krispy Kreme doughnuts, kettle chips, and Kit-Kats, making them your official comfort foods.

Lastly and importantly, processed foods affect a part called the *insula*, which is responsible for interoception, or the ability to interpret your body's inner signals. This is a crucial element of the gut feeling we delved into a little earlier, allowing you to feel like yourself from the inside out. You can easily remember the name insula just by associating it with the word *intuition*, since the insula translates signals that arise from our guts and turn them into subjective experiences, which include intuition and bodily awareness. Chronic consumption of ultra-processed foods muddies this clear connection, blunting the awareness you would otherwise feel and a clear sense of knowing what your body actually needs. One example is a feeling of fullness, which many would argue is subjective. However, a well-tuned insula would make you stop eating before you're overly full, resulting in that nice, satisfied feeling and not a sickly, heavy one. Another example would be a gut feeling about someone. That flicker in your stomach when you receive a text from someone who does nothing but stress you out is a sharp insula-related signal that indicates emotional awareness, a strong 'off' feeling about something or someone.

However, unlike most other nutritionists, dietitians, and holistic practitioners out there, I'm not going to tell you to stop eating processed foods completely. Yes, you heard that part right. After restrictive eating patterns in my teens and early twenties, processed foods had an important part to play in my recovery, and I'm grateful for them. This is why I still advocate the regular cafe pit stop for your flat white and pastry soul-soother. Food is absolutely essential to not just life, but a beaming social life. I grew up in a food-obsessed family, with my maternal grandmother whipping up Chinese feasts every week, and until this day, we try different restaurants alongside old favourites to supplement our Wednesday night dinners and Sunday afternoon family lunches. I enjoy meeting my best friends over coffee, and my husband and I relish the same coffee and perfectly toasted ham-and-cheese toast that's filled generously with a variety of cheese and jam hauled all the way from Brazil. I knew I fell in love the moment I realised he enjoyed cooking good food during our spare moments together, sourcing out new and exotic ingredients I would not otherwise try alone. Good food and the craftsmanship surrounding it should be celebrated, not devil-labelled.

Radical Mindfulness

Having discussed and personally acted on the best methods for incorporating sustainable, healthy habits around any type of food with nutritionists and holistic practitioners alike, what I'm bent on encouraging is not the elimination of these foods, but the management system behind them. It is very hard to control your social calendar, with the random work lunch, client dinner, or family outing, but there is a way to reset your internal system to implement calm and regulation. The approach I return to again and again is what I call *radical mindfulness*, a behavioural elegance that makes you impervious to your surroundings and restores clarity in mind and body. It is a system that works to mirror your motivations and actions with your inner self, a self imbued with a painfully sharp awareness of your body and the clear signalling between your gut and your brain. When that alignment is restored, you become less reactive, more attentive, and start to love food again.

This occurs with three core things that happen at and after every mealtime: *excruciating slowness*, *portion calibration*, and *delayed integration*.

Sounds simple, doesn't it? These are all words and techniques you may have heard of before. However, what's missing is implementation and the understanding that there is no diet formula beyond order, timing, and attention. It is a recipe that will retrain your stomach to handle the amounts of food you were born to handle, not stretched out so badly that it blurs the signals you need most.

Starting with the first and most important point, *excruciating slowness*, I call it radical because this is genuinely transformative—you are effectively reclaiming control over your internal state, thus not letting the external environment control the way you behave around food. You are not going to finish your plate just because it's there, or eat and drink just because others are doing so, or reach for food out of boredom or to regulate a feeling that you haven't quite placed your finger on yet. This is a core behaviour technique because it bleeds into all aspects of your life, extending far beyond the dinner table. You know the phrase 'the way you do anything is how you do everything'? It rings true here. When you learn how to take a deep breath, pause, and really relish the food you eat at the table, you will do it everywhere else: before reacting to an angry mother's text, before impulsively putting down a down payment on a property, before spiralling emotionally in an argument with your partner. Eating slowly, chewing until food is mush, allowing meals to take at least twenty minutes, whether it's an almond croissant or a salad, gives your nervous system time to register safety and fullness. Yes, it takes twenty whole minutes for your stomach to properly register when it is full, and most people are used to wolfing something down in five—it's no wonder that five minutes later, you still feel hungry and need a snack or coffee to tide you over into the afternoon.

This shift from fast to slow reduces stress-induced water retention, bloating, and puffiness—outcomes many women experience when they choose restriction instead of regulation. You will watch your mind soften and weight drop off seamlessly. When you drastically slow down and treat your food with care and attention, you regulate your nervous system and remind your brain that you are not operating under threat.

If you have a history of binge or disordered eating or frequently eating very large meals generally, then yes, the signalling between your stomach receptors and brain to tell you to stop eating are blurred; however, it is only a matter of weeks or days for the main 'I'm full' hormones (called GLP-1, PYY, and CCK) to kick in and let you know to stop eating. Your stomach literally shrinks, which means the mechanical stretch needed to induce that feeling of fullness also comes back. The body is forgiving, and things will return to where they should be, and I know because I have lived through it.

During my forced feeding days, I bulldozed through piles of food just to get my weight up, and it obviously worked, for I managed to gain ten kilograms in the space of a few years. However, I didn't feel bloated, distended, or generally uncomfortable. Only when I returned to structured mealtimes, understanding portions, and slowness did the excess weight fall away, quietly and permanently. When you realise you may be sliding back into old overeating habits, as I did when I returned to Singapore, going back to excruciating slowness always works: take a deep breath, slow down, meditate on all the wonderful tastes.

Have you ever had a meal that wasn't so heavy, but you noticed hours later, that you're not jumping up and down for dinner? *Delayed integration* is the practice of allowing the body time to respond before drawing conclusions. Rather than evaluating a meal, a habit, or a choice in the moment, you check in hours, not just minutes later. What you're trying to keenly notice is, do you feel clear and light or heavy and foggy, a general steadiness or agitation/irritation in your gut? This matters because many of the signals that shape mood, appetite, and energy do not resolve immediately, and even more so if you have not had the best relationship with food in the past. Blood sugar regulation, inflammatory signalling, neurotransmitter balance, and nervous-system tone all unfold over time.

When we assess too quickly, we misread the body. We assume something worked or didn't work based on immediate sensation, and in doing so, we override the slower, more accurate feedback the body is trying to give us. Delayed integration restores that accuracy by shifting decision-making away from impulse and towards pattern recognition.

Importantly, you begin to notice which foods and meals leave you feeling a certain way. It could be good to start journalling about it. How did that brunch with friends earlier make me feel? Why does eating meat always leave me feeling heavy, not energised? Food purity (healthy or unhealthy) does *not* matter here. No meal is labelled good or bad; it is simply informative. Take a Black Forest Christmas-ready trifle, for example. It is consumed slowly over the evening, with a few spoonfuls here and there, a pause for conversation, and sips of Sauternes in between. The body has time to register sweetness, fat, and fullness without urgency. Nothing spikes, nothing crashes. Hours later, you are satisfied and strangely calm. Contrast that with a 'clean' meal eaten in a rush—a virtuous salad inhaled between emails or on the plane. Despite its nutritional profile, perhaps it leaves you feeling a little unsettled or hungry again shortly after. Delayed integration makes this visible. It's not the ingredient list that determines whether a meal grounds you, but the pace, attention, and nervous-system state in which it's consumed.

Delayed integration helps you trust your body and choices again—not just in its signals, but in your ability to respond to them calmly and intelligently. It is what turns radical mindfulness into a learning system rather than a performance. All this tuning in to how you feel and constantly assessing your response sounds may sound slightly overwhelming, but behaviour shift lags behind awareness, changing gradually as the feedback integrates naturally, rather than through conscious rule-setting.

Now that we have anchored the importance of excruciating slowness and delayed integration, the next points naturally follow. They are not separate techniques, rather extensions of the same intentionality and clarity we want to cultivate fully. They work hand in hand with radical mindfulness to reinforce your gut–brain dialogue at every meal.

In the same vein, understand the power of *portion calibration*. When you're with other people with large servings of food in front of you, make sure to have a plate in front of you. Whatever the food is, whether that be steak or nachos, get a bit of everything and then shift your focus properly outwards, to connection and conversation and presence. If you're alone, continue tuning in to feel and taste what's going on in your mouth.

Fill your plate with intention instead of randomly picking at things. Remember, this is not a manual dictating 'how much' or 'how often', but an invitation to tune inwards, not outwards. This subtle shift from fixation to outward focus prevents you from slipping into reactive eating, anchoring you back into choice, the choice of having total and absolute control over what lives inside your body each mealtime. A key helpful habit here is to prime your gut with protein first. Protein kick-starts the fullness signalling, reduces cravings, and keeps your mood stable. It is the raw material needed for the construction of neurotransmitters such as dopamine and serotonin, and the very lining of your gut, making protein intake foundational not just for appetite control, but for digestive resilience and emotional control. Even if you do have dessert or anything sweet after your meal, you won't be suffering those intense glucose spikes, which cause your mood and emotions to go haywire, making you lurch forward to grab second, third, and fourth helpings. From a neurological point of view, protein is also needed to prevent a hijacking of our reward system circuitry, that cluster of brain areas mentioned earlier, including the nucleus accumbens and ventral segmental area, areas that were found to light up even more when protein was restricted in humans and animals alike.

There is no shame in taking food away from a restaurant or leaving something behind. If you, like me, grew up with a no-waste mindset, thanks to my father, who insisted that I would sprout acne with every plate unfinished, it may take some time to disentangle the notion that leaving something behind is acceptable. Releasing this narrative releases a scarcity mindset. Food is always there; the fifth bite tastes the same as the first couple. It is no longer something you must earn or cling to desperately. Unlike our hunter-gatherer ancestors, we have no need to fear the sudden stealing of food by predators or enemies. Three bites of that almond croissant will taste the same as the fifth, so once you're satisfied, you can keep the rest for later or simply leave it if you don't yet trust yourself with the tools to not touch food. *Food will always be there.* It is that simple.

Eating out of boredom or habit, like smoking, chewing gum, or vaping, is indicative of oral fixation, the foundation of uncontrolled eating habits. The rhythmic chewing and puffing are soothing because it activates the vagal pathways, which feel too good for you to want to stop. Immediate oral stimulation also provides this false sense of relief or control, since it floods you with dopamine and keeps you distracted from what's going on and how you feel within your body. As much as good food cannot be missed, it fundamentally distracts from the interoception and intuition we are trying to cultivate in our journey to optimise our physical and emotional health—to manage or drop weight effortlessly, feel great all the time, and have the tools to sustain our healthy habits. Therefore, it is essential to learn how to go inwards before giving yourself full permission to enjoy all the good food the world has to offer, and you do that by cultivating radical mindfulness: drastically slowing down to properly taste your food, taking at least twenty minutes to chew and finish it, calibrating your portions with intention, tuning into your internal cues with clarity.

The body does not detox by withholding or taking away things, but it does so via regulation. Once this cornerstone habit is mastered through the aforementioned techniques, you're good to go. Every meal is a reminder to seriously slow down, pay attention to how much you are consuming, and listen to what your gut is saying, the whole way through. After the first two to three days, stretch receptors in your stomach will start to recalibrate and thus shrink that distended stomach (if you have been chronically overeating or had a history of disordered eating in the past), and vagal tone increases through slowing down. By the second week, inflammation markers will drop, GLP-1 and PYY response normalises to regulate your appetite, and your general urge to go crazy and binge on everything in sight will subside. You experience clarity of mind and a stabler mood.

Once you have gone through physical and mental reclamation of your body after these few weeks, you gain the kind of power that feels effortless. From here, you can introduce whatever foods your heart and soul desire, because you are in the proper mental framework to slow down and enjoy what's in front of you without losing yourself in it.

Mood-Altering Nutrients

Now that the behavioural architecture is in place (slowness, delayed integration, and portion calibration), we can turn our attention to the *biochemistry* beneath it. Behaviour sets the stage, but nutrients do the heavy lifting inside your gut–brain axis. However, without understanding what's happening at a neurochemical level, even the most elegant habits can only take us so far.

This distinction does matter to me personally. As a teenager, I lived for years in a restrictive subtype of eating that, at the time, felt strangely functional. I didn't feel overtly distressed—in fact, I often felt just fine, but it was only later through neuroscience that I understood why. Neuroimaging shows that in restrictive eaters, the brain's reward circuits, which are the same ones that light up for food, love, and joy, begin to dim over time. The prefrontal cortex, which is the part of the brain responsible for control, overpowers the limbic system, which governs emotion and appetite. In other words, the brain learns to prize control over pleasure. No wonder I 'felt fine' as a teen—my brain literally was reprogrammed into enjoying the feeling of control and lightness because it sounded out the areas of my brain that prioritised how I felt and blurred my normal hunger signals too. I remember staring at an fMRI image in one of the studies, which was literally a map of self-denial, and realising I had lived that scan. I had trained my brain to find safety in suppression, to seek comfort not in nourishment, but in this illusion of order and control. The science mirrored the ache I had felt all along: that starvation doesn't just numb the body; it numbs the self.

What struck me most was that this pattern wasn't unique to me. The neurobiology of restriction, or that quiet shift towards control and emotional narrowing, echoed themes I would later encounter again, this time not in myself, but in the data. What began as a purely personal reckoning gradually evolved into an academic question: if mood, reward, and regulation could be reshaped so profoundly by deprivation, what other biological systems might be quietly shaping how we feel?

During my doctoral studies, that question took a more concrete form. I came across something unexpected: people with depression often have lower levels of certain immune molecules called interleukins, which typically help the body respond to stress, infection, or injury. Studies have shown that depression is often associated with immune system changes, including altered levels of interleukins like IL-6 and IL-1β, which are involved in inflammatory responses. [1] I noticed a similar pattern with lipid metabolism, or *types of fat in the body*, particularly HDL and LDL cholesterol (often called good and bad cholesterol), which are responsible for transporting fats through the bloodstream.

At first, this was puzzling. Depression is frequently framed as an inflammatory condition, and inflammation usually implies immune activation, not suppression. So why were these signals elevated, not reduced?

These inconsistencies raised new questions: *Could mood disorders be connected to imbalances in the immune system, metabolism, and signalling?* And more importantly, could the gut (which sits at the intersection of both immunity and metabolism) [2] be playing a much bigger role than we ever thought here?

To understand how the gut exerts this profound influence on our bodies, we need to look closely at the actual nutrients that directly shape mood, cognition, and emotional regulation. As discussed earlier in the section on radical mindfulness, it is key to remember that *how* you eat is just as important as what you eat, if not more important. However, these compounds are worth noting since they don't just fuel the body; rather, they act as signals within the gut–brain axis, really fine-tuning and clarifying the gut–brain signalling we aim for, influencing how we feel and function every day:

Tryptophan: A special amino acid found in eggs, chicken (particularly the breast meat), turkey, tofu, and seeds, it's a precursor to serotonin—that neurotransmitter best associated with mood and emotional regulation, and crucially, most of it is made in your gut, not the brain. However, tryptophan's impact on mood is not determined by intake alone. What matters just as much is how the gut processes it. In a well-functioning system, specific gut bacteria help direct tryptophan down the serotonin pathway, but when microbial balance is disrupted, tryptophan is likely to be shuttled down other metabolic routes that have little to do with optimising your mood. In these cases, the signal just doesn't land. In other words, mounds of chicken breasts won't do much unless your gut is equipped to process them.

Omega-3 fatty acids: So handy, very delicious, and probably the most studied nutrient of all time, especially in relation to brain health. These anti-inflammatory fats are critical for nerve cell fluidity and brain function. Found in fatty fish, walnuts, and flaxseeds, they help balance out the inflammatory effects of omega-6 fats common in processed foods. They also influence neurotransmitter signalling, including that for serotonin and dopamine, by affecting how efficiently receptors respond to these chemical messengers. During my doctorate, which examined how metabolites, nutrients, and lifestyle factors relate to susceptibility to major depressive disorder, one of the most striking and consistent findings was that lower circulating levels of lipids, or specifically fatty acid subtypes, were associated with a higher risk of depression. [24] This was initially counterintuitive, given how strongly fats have been demonised in nutritional discourse over the past few decades. When is the last time you heard about someone being firm about low-fat everything? Probably just this morning at your local cafe. Yet when viewed biologically, the relationship makes sense. The brain is composed largely of fat, and neuronal communication depends on our precious lipid-rich membranes functioning smoothly. When omega-3 availability is low, cell membranes become more rigid, inflammatory signalling increases, and neural signalling becomes less efficient. All of this can contribute to low mood, cognitive slowing, and emotional blunting. Adequate omega-3 literally impacts the physical architecture of the brain, therefore affecting how we think, feel, and move through our lives.

Fermented foods: Yogurt, kimchi, kefir, miso, and yes, the occasional glass of wine. These contain the live bacteria our guts need to reinforce microbial diversity. A diverse microbiome tends to be more resilient, better equipped to produce neurotransmitters, and more effective at regulating inflammation when stress and obstacles hit us in the face. It's even been found that fermented dairy, such as Greek yogurt, skyr (the thicker, high-protein sister of yogurt), and kefir in particular, can help repair a damaged gut lining, which is essential to first restore even before trying to fix other things such as the amount of protein and fat in your diet, [25] because if your gut lining is damaged, even the best foods can slip past without doing their job. The reason is simple: if you're not absorbing these things properly, what's the point?

Polyphenols: These are important chemicals found in berries, green tea, olive oil, and dark chocolate. Polyphenol-containing foods are easy to spot at the grocery store, since they're usually very pigmented. The brighter the food, the louder the message it sends to your gut. Think juicy and bright berries (intense blue), spinach and kale (deep green), red onion and beets (bright red). Green tea or matcha is a personal favourite, since you can work it into your day without the highs and crashes you get with normal caffeinated coffee. These antioxidant compounds are metabolised by your gut microbes into beneficial molecules that improve both gut and brain health. [26]

Glutamine: An underrated amino acid, glutamine is the precursor to GABA and glutamate, which are the brain's primary inhibitory and excitatory neurotransmitters, which together regulate alertness, stress responsiveness, and a general feeling of calm. Just as importantly, glutamine is a key fuel source for the cells that line our gut, consuming it more than any other organ in the body. This is why during periods of physical or psychological stress, the body's demand for glutamine increases, and it's used faster than it can be produced, so it is always handy to consume external sources of glutamine to ensure that it is ready and available, helping your gut remain resilient and intact. Obtaining glutamine from dietary sources such as chicken, eggs, tofu, and dairy products can make it easy for you to support these systems, especially during times of heightened stress. [27, 28] In some cases, getting glutamine via supplementation may be useful as a short-term support while gut integrity and its overall resilience are being restored.

Collagen: Did you know that after the age of 25, your natural collagen production starts to decline, which is a gradual but meaningful feature of aging? [29] It sounds like bad news, but it's one of those ageing facts you can do something about. Collagen is an essential component of your hair, nails, and skin, so it helps with everything from reducing hair and nail brittleness to wrinkle depth. It is particularly relevant to gut health, since it provides the structural architecture of your intestinal lining. A healthily lined and nourished gut reduces the unwanted permeability that happens without it and thus leads to background immune activation that heavily interferes with everything from your digestion to mood and metabolic regulation. In this way, collagen supports your aesthetics by first supporting the system responsible for absorbing and utilising nutrients efficiently. You can think about collagen like the mortar between bricks—you may have a solid mound of bricks to start with, but without the gluey lining to hold it all together, even the most solid bricks become porous and unstable. You can find collagen as a powder or in bone broth, another personal favourite, and best of all, you can mix it into things like smoothies and even hot chocolate, so you get a protein-filled snack or coffee without the glucose spike from a purely sugar-laden afternoon snack. Growing up, I never gave much thought to the clear fish soups and broths that accompanied dry noodle dishes characteristic of Chinese cuisine and tradition, yet these were quietly rich in collagen, antioxidants, and nutrients that later proved surprisingly hard (or expensive) to replicate when I moved to Britain. I have a newfound appreciation for these dishes, which I now consume frequently. Now, I personally enjoy mixing collagen into my mid-afternoon matcha or simply shaken with water—it's a small tweak that feels like care from the inside out.

Bile acids: Crucial yet rarely discussed in the context of gut health or otherwise, bile acids sit at a critical intersection between digestion, metabolism, and mood. Produced in the liver and released into the gut, bile acids are best known for their role in fat digestion, but they also function as signalling molecules, influencing inflammation, energy regulation, and gut–brain communication. When bile flow is sluggish or dysregulated, fats linger in the gut, microbial balance shifts, and meals can leave you bloated, foggy, or strangely flat. Over time, this can contribute to low-grade inflammation and impaired gut signalling, with downstream effects on your mood and metabolic regulation. This is also why I loathe the low-fat-everything epidemic, which has infiltrated and somehow lingered throughout society—you are doing little for your body and mind by taking it away, and usually low-fat products replace the fat with sugar and other chemicals, which only worsen your physical and emotional status anyway. This, of course, supports the doctorate work I did, which showed the significant relationship between low lipid levels and low mood—health should not be joyless. Chronic fat avoidance can further weaken bile production, creating a loop of intolerance that is often mistaken for sensitivity. Fats must be celebrated and enjoyed regularly, and when consuming the right fats to support bile acid production, you will feel grounded instead of overwhelmed. Fats to stimulate bile release are olive oil, butter, avocado, eggs, fatty fish, and it would also help to prioritise choline-rich foods, since choline is needed to make the main component of bile and thus support that precious gut–brain signalling. These include egg-yolks, liver, seafood, and soy. Most important, maintaining (more or less) regular meal timing helps plenty here, as bile responds best to rhythm rather than erratic intake.

The Sugar-and-Stress Tango

Sugar deserves its own subsection, especially as I'm a self-confessed former sugar addict. While a bite of chocolate can soothe in the short term, excessive sugar intake can be emotionally destabilising in the long run. This is particularly relevant to women. For us females, there are sharp fluctuations in oestrogen and progesterone around our menstrual cycles, which influence serotonin and GABA levels respectively, hormones tightly linked to our sense of reward and mood. [30] Combined with a generally heightened stress reactivity, this means we are more prone to sugar cravings than men.

Feeding your sugar cravings during your period feels great in the moment, and in measured doses, it isn't altogether a bad thing, but the issue arises when sugar becomes your default self-soother and emotional regulator. This habit unfortunately feeds pro-inflammatory bacteria, destabilises your blood sugar, and impairs the gut lining—all of which lead to emotional rollercoasters alongside metabolic havoc. Behind the scenes, sugar also nudges your insulin (hormone that manages your blood sugar) and cortisol (our main stress hormone) into overdrive. Over time, this dulls your dopamine response (the system that says 'Hey, life feels good'), leaving you foggy, flat, and wondering why everything suddenly feels a bit grey.

Is the hunger in your gut or your head? That was the question I started asking myself during late-night snack raids during graduate school—an occasion far from rare. Was I truly hungry, or was something else driving the urge? Not all cravings are yours; some are seeded by your gut bugs. Yes, some of these microbes thrive on sugar and can influence cravings accordingly, nudging you towards the chocolate-chip cookies that sustain them. [31] In that sense, cravings can reflect microbial demand rather than physiological need.

This is your rewire shift: recognising the difference between physical hunger, emotional hunger, and microbial signalling. When your gut is balanced, your cravings soften and stabilise. When it isn't, it may feel like your cravings are chaotic and impulsive, as though you need the extra bites long after you're satiated at dinnertime.

The empowering bit? You can take back the wheel by going back to those basics, which are handy regulatory tools to always come back to and not strict rules: protein before sugar or at least combined with it (think sausage with a pancake, not the pancake first), and drastically slowing down and eating with presence to restore your agency and reduce urgency.

Rebuild a Calmer Relationship with Food

Healing the gut isn't about restriction. It's about recalibration (rewiring) and regulation. The whole premise of *Gutsy* is to zoom in on our guts being the underrated mediators of energy and cognitive parameters you may not have thought they would affect, such as your mood and relationship with food itself, about clarifying your internal state so that the rest follows suit. Without any sort of recalibration, there is no clarification of the signalling between your gut and brain, and you will never understand why you cannot achieve physical and emotional harmony. Instead of crash diets and food rules, think nourishment and keep going back to those pillars in radical mindfulness. This is where you naturally become incredibly mindful of every bite you take, and eating stops being automatic and is much more informative.

Tuning back into your hunger, your fullness, your fog or focus, is self-respect in action. Emotional eating or eating out of sheer boredom isn't failure; it's usually your gut trying to get your attention. This is the work I do with every client, and it's what I had to confront myself with.

Quick Dietary Tips for Your Mood and Physical Health

Begin each mealtime with protein and fat, not sugar or straight carbohydrates. A muffin and a latte first thing in the morning is probably one of the worst things you can do to your body metabolically. If you just can't give it up, start by having half a muffin with eggs or yoghurt. Even this small addition of protein raises the glucose spike by at least 20–50 per cent, and as the days go by, your muffin-and-latte craving will wane. In a similar vein,

Don't demonise carbs, but pair them with fat or protein to stabilise your blood glucose levels and mood. Have your avocado toast, but add an egg or two on top. This means cooking and enjoying the pasta, but pairing it with a side salad and a handful of protein as well. This works well with the portion calibration component of my radical mindfulness piece explored, where you would do well to make sure to plate everything—a palm-sized amount of protein, then fill the rest of your plate with carbohydrates and fibre before slowly enjoying it. Pleasure stays; volatility reduces.

Drink coffee with or after a meal, not before. Have a decaffeinated coffee if you really need something warm and ritualistic to enjoy first thing in the morning, like me. As much as I adore coffee, you really don't need the spike of cortisol (and thus stress) first thing in the morning, adding noise to the gut–brain connection we want to restore and protect.

Practice radical mindfulness. Keep bringing it back to this one. Engage your senses. Let meals take time and not give in to eating hurriedly just because everyone else seems to be. Seriously enjoy your food, give thanks for it, make it a meditation. Presence will recalibrate your portions naturally: the fifth bite rarely holds the same value as the first few, and you begin to realise you need *much* less than you think. How you digest food is how you digest life; rushing one means you end up skimming through the other too.

Include fermented foods daily, including fermented dairy, even in small amounts. These repair your gut lining and thus the signalling between the gut and brain, improving clarity and emotional stability.

Hydrate, especially if you drink caffeine, optionally with electrolytes. Most of us drink water but miss out on deeper hydration. Electrolytes such as sodium, potassium, and magnesium are essential for neuronal firing, vagal signalling, and gut motility, all of which shape how clearly the gut and brain communicate. When electrolytes are depleted, the body often mislabels thirst or mineral deficiency as hunger, leading to unnecessary snacking and blood-sugar fluctuations that further cloud gut–brain feedback. So a lot of the time, you think you need food, but really, you're just missing out on some key nutrients. I usually have a scoop of flavoured electrolytes in water one to two times a day—delicious, fun to drink, and ideal to prime your gut before it faces caffeine and food.

Prioritise omega-3s. Reduce neuroinflammation by consuming more fatty fish like salmon, and incorporate flaxseeds and chia seeds into your diet at least two to three times a week. These suppress your body's stress response, keeping you calm and cool throughout the day. Of course, omega-3 supplements do make this easier.

Use turmeric and ginger in your cooking. These are serious anti-inflammatory and antimicrobial powerhouses, which will calm your gut and thus the rest of your body too. They support the gut environment that underpins clear gut–brain communication—both contain bioactive compounds that help dampen low-grade gut irritation and shape microbial balance, reducing inflammatory noise that can distort how bodily signals are perceived. Calm the gut, clarify the signal.

And the last and perhaps most important point: *There is no such thing as bad food.* Once the gut has been recalibrated with a period of time implementing the above dietary tips and nutrients, you can adjust your lifestyle to eat the tiramisu and have the burger. Enjoy food fully—especially when you travel or try a new restaurant or cafe, because novelty and pleasure are what make being alive so exciting and beautiful. What matters more than the novelty, however, is pace, presence, and pairing foods in a way that won't put you in a metabolically chaotic spiral. We have evolved beyond grazing like animals for good reason. In this chapter, we've come full circle: from understanding how your gut affects your feelings, to now seeing how your daily food choices directly impact that gut–brain connection. Food isn't just a lifestyle choice. It's a tool, a therapy, and when used right, a love language to your nervous system. Feed your nervous system, and it will feed your soul in return.

End-of-Chapter Rewire: *Which specific food pattern or nutrient strategy will you intentionally swap or incorporate this week to tune your biology for better mood and energy?* Choose one. Your brain and gut will notice the difference.

Next, we'll zoom deeper into the molecular consequences of gut imbalance and what happens when gut communication goes awry at the level of neurotransmitters and inflammation.

CHAPTER 4
MOLECULES THAT MATTER (THE REWIRE TIER)

Why Metabolites Deserve Your Attention

Back when I was deep in my Ph.D., I spent most of my days buried in papers about molecules called metabolites: tiny by-products of metabolism, produced as the body breaks down food and manages energy or stress. Call them biochemical leftovers all you want, but metabolites tell you plenty about the state your body is in.

One of the more interesting discoveries was during my execution of a large systematic literature review, where I found the presence of elevated levels of two metabolites in people with major depressive disorder: the amino acid *alanine* and *citrate*, a key intermediate in the Krebs cycle, or the body's main energy generation pathway. [32] My supervisor and I formed a hunch: What if these unusual high levels pointed to something deeper—like the body trying to patch up a glitch in its energy system? The eventual hypothesis was that the body might be pulling alanine from muscle tissue to keep things going, and low levels of glutamate that corresponded to low mood were leading to an accumulation of citrate. These patterns pointed to metabolic stress overall, not just isolated abnormalities. I started to connect the dots between mood disorders and metabolic health, which continued to be a growing interest alongside my research at Oxford.

This was the moment I began connecting mood disorders to gut health in a more serious way. What if these metabolic disruptions weren't originating in the brain at all? What if they were downstream effects of gut imbalance, where microbial dysregulation altered nutrient processing and thus also metabolite production? [33]

When Gut Imbalance Becomes a Metabolic Problem

This brings us now to the fourth and final knowledge pillar of this book—that is, gut dysbiosis disrupts the production and regulation of essential metabolites, with far-reaching consequences for metabolic, physical, and mental health. [34]

Metabolites are by no means passive by-products. Many are produced, modified, or regulated by gut microbes, and they fall into a few broad, overlapping categories. Here are five categories:

1. *Signalling metabolites.* These are needed to regulate brain function (including our mood), immune responses, and gut motility. These include hormones such as serotonin and dopamine, bile acids, and some amino acid metabolites, such as tryptophan derivatives.

2. *Energy and fuel metabolites.* These comprise the SCFAs we mentioned earlier, as well as branched-chain amino acids, or BCAAs. You may have come across this term in an exercise or physiology context since it is known to reduce muscle soreness and fatigue by stimulating your cells to produce more energy. [35]

3. *Immune-modulating metabolites.* Polyamines and lipid metabolites support the integrity of your gut lining and modulate inflammation to support overall gut health.

4. *Antioxidant metabolites.* These are also known as flavonoids and phenolic compounds, which buffer inflammation and oxidative stress.

What is striking is that gut-derived metabolites overlap with all these systems at the same time. They help determine how efficiently you generate energy, how reactive your immune system becomes, and how clearly your brain interprets signals from the body. [36] When this system is disrupted, the effects rarely stay local. Inflammatory signalling can become systemic, interfering with communication in your brain and thus affecting everything from how clearly you think (hello, brain fog) to emotional processing (hello, mood swings). In other words, what starts in the gut can quietly affect how you feel and function.

Take serotonin, for example. When you eat things that contain the chemical tryptophan, such as chicken, eggs, or tofu, it's metabolised by our gut bacteria to produce serotonin. Serotonin itself cannot directly cross the blood–brain barrier (BBB), the selective barrier that protects our brains. Not everything can pass through, and serotonin is too bulky and lacks a transporter that can help it move through this barrier. However, serotonin levels in our gut can indirectly influence levels in our brains, because of signalling via the vagus nerve (I will add a note here saying that serotonin can be made in our brains, specifically the cells of our brains called neurons, since tryptophan can pass through the BBB and be converted locally).

Specifically looking at serotonin, aka our happy molecule, it's important to note that it does not always follow this path of being metabolised from tryptophan. Suppose our guts are not in balance, aka in a state of dysbiosis. In that case, tryptophan is metabolised via what's called the *kynurenine* pathway, which means it is broken down into these chemicals called kynurenine and quinolinic acid, instead of into serotonin. [37] This means less production of the happy hormone we want more of and a higher propensity for depressive symptoms, which characterise mood disorders such as major depressive disorder and generalised anxiety disorder. The diagram below shows the simple but powerful upregulation of this kynurenine pathway when you have gut dysbiosis and a lot of inflammation ravaging your gut and whole body. [38] This upregulates the production of more pro-inflammatory cytokines, creating this positive feedback loop, which only increases inflammation locally in our guts and systemically in our bodies, wreaking havoc everywhere [39–42].

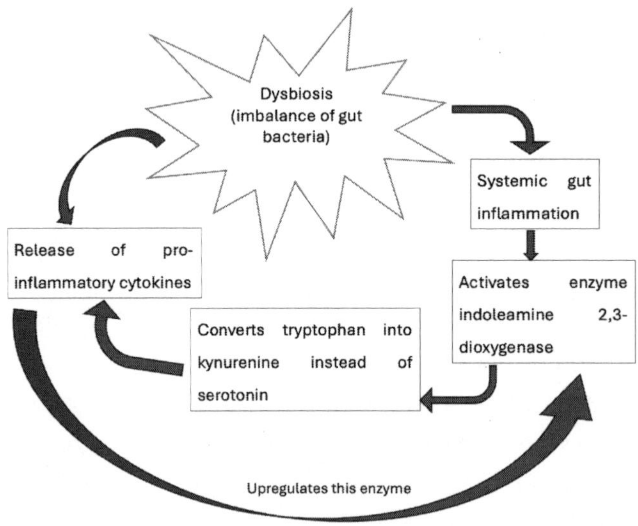

Figure 1: Dysbiosis mediates the conversion of tryptophan to kynurenine instead of serotonin

Rewire Shift: From Chemistry to Agency

This is where it becomes actionable. If your gut is pushing tryptophan down the kynurenine route instead of making serotonin, you're not just dealing with a digestive issue; you're facing a mood-altering metabolic switch. Restoring balance to your gut could mean reclaiming your ability to produce and regulate your happy hormone. This shift doesn't just ease symptoms; it transforms the root of them.

You might wonder: if our brains can make serotonin, why does gut serotonin matter? Well, your gut actually produces about 80 per cent of the body's serotonin, and while it doesn't all reach your brain directly, it plays a massive role in mood, immune regulation, and reducing inflammation. And when inflammation is up, the brain gets less of the tryptophan it needs to make serotonin on its own.

End-of-Chapter Rewire: *Considering the molecules that matter, what is one new intentional habit you will adopt to support your gut's molecular communication and reduce inflammation?*

Think small and repeatable—a food, a walk, or a mindset rewire that supports your chemistry.

You've begun reshaping how you eat, move, and respond, and are more educated on the gut–brain axis and why our guts play such a significant role in our lives. The rewiring has started. But real transformation goes beyond daily routines; it's about who you become when you live in sync with your system. It's now time to reinvent.

In the final chapter, I'll guide you through how to support your gut so it can support you and specifically you, from the foods that build a healthy microbiome to simple lifestyle shifts that can calm inflammation and boost your energy.

TIER 3
REINVENT

CRAFTING YOUR PERSONALISED GUT HEALTH BLUEPRINT

We have arrived at the practical and moving part of this whole book. *Reinvent* is about using personalised gut–brain health strategies to stop guessing and start healing. Instead of following generic health advice, you'll learn how to work with your body's biology to get real results. This chapter shares how tracking gut function, adjusting food timing, and responding to metabolic feedback can help you feel clearer, lighter, and more emotionally stable. It's not about doing more—it's about doing what works for you. Through tests, data, and practical tweaks, you'll discover a system that honours your individuality, rebuilds your gut, and supports your physical and mental transformation.

Key Words

SCFA (*short-chain fatty acids*): Fatty acids produced by gut bacteria from fibre; support gut lining and brain health.

GABA (*gamma-aminobutyric acid*): A calming brain chemical produced by the gut that helps reduce anxiety.

Tryptophan: An amino acid used by the body to produce serotonin, a mood-regulating hormone.

Personalised gut report: A microbiome test that analyses your unique gut bacteria to guide diet and supplement choices.

Gut–brain axis: The communication network linking your digestive system with your brain, affecting mood, memory, and behaviour.

Zone 2 exercise: A moderate-intensity exercise zone (akin to a brisk walk, usually) that improves fat burning and mitochondrial function while reducing inflammation.

Neurotransmitters: Chemicals like serotonin and dopamine that transmit signals in the brain and influence mood, focus, and energy.

CHAPTER 5
PERSONALISED GUT-BRAIN SOLUTIONS TO OPTIMISE INDIVIDUAL HEALTH

Graduate school saw me tracking everything: my workouts, macros, mood, menstrual cycle, and yes, even my bowel movements. On paper, I was doing everything right: I believed I was choosing the right foods, moving daily, stayed relatively disciplined. Yet something didn't feel quite right. Deep down, I knew my relationship with food and my overall health could be improved: I was rushing through meals instead of really enjoying or savouring them, and thought I was eating well during the day, but somehow at night, I couldn't keep my hands off sugar. It felt sometimes as if a robot was working within me, pulling my hands and legs to the kitchen when I had no business being there, and the day was sealed with a distended gut and numb mind. The nourishment of daily walks and salads I painstakingly put together was negated by a mind not yet primed to receive and digest food the way it should; a constant ebbing undercurrent of low-grade stress and urgency bled into every meal, blunting digestion, absorption, and the healthy mind–body signalling I craved. At the same time, my parents' divorce, change of partner, and change of environment created a sense of unease, destabilising my sense of control and safety, which then manifested in a strange relationship with food and my body.

Funny how you can Google all the wellness tools under the sun, yet not quite know the perfect formula for *you.* I'd wake up most mornings looking two inches larger from all angles because of water retention. My head felt slow and heavy; it got to a point where my supervisor noticed and suggested giving me a break. It wasn't burnout; it was brain fog preventing me from wanting to even get out of bed most days.

Nothing made sense. I was keeping it healthy in all possible ways—eating well and right, moving every day, but every step was a drag. It was only when I introduced and tested the implementation of radical mindfulness and later paired it with comprehensive gut health and metabolic testing that the pattern began to make sense. What was missing was a sense of rhythm and readiness. Sure, I thought I was taking care of my gut health, but was consuming regardless of whether my body was primed to receive, and eating in states where digestion simply could not do its job, leaving me feeling puffy, in a body that seemed to cling to every morsel of food, making effortless weight management near impossible.

Radical mindfulness taught me, and will teach you, how to best listen and thus react to bodily cues. This, combined with restoring rhythm with sleep, pausing throughout the day, and regular meals eaten with a receptive nervous system, will actually move the needle.

When the data finally matched the feeling

Testing was rather revealing. Not in a dramatic way, but in a quieter, confirming way that made everything make sense. Sort of like experiencing chronic, hard bloatedness and nausea, so you take a test and realise you're pregnant—you're no longer gaslighting yourself. Blood tests revealed that my fasting insulin was higher than it should have been for someone eating well, pointing to poor glucose handling beneath the surface, since insulin is what shuffles glucose from blood into cells, and elevated levels show that your body is working harder than it should to do so. Homocysteine and C-reactive protein levels were also higher than they should be, suggesting that despite a nutrient-dense diet, my body was struggling with gut malabsorption and a surprisingly high inflammatory load—I was stressed out, and it definitely showed. Now on to the more unique profile brought on by my own research that found a significant inverse relationship between lipid levels and mood; overall, lower lipid levels are strongly linked to low mood, which is counterintuitive to mainstream thinking that high-fat diets are inflammatory and bad for you. My own lipid profile—particularly circulating lipid levels—was lower than expected, echoing this pattern of low lipids to metabolic dysregulation instead of just cardiovascular health, which is what lipids are almost always associated with. I was concerned but weirdly reassured when this was all done in combination with testing my gut and its health: my body was in defence mode and only wanted to protect me after years of restriction and confusion with no real routine, and this gave me something to work with and correct.

Individually, none of these markers would have raised alarm bells, but together, they told a coherent story. This body was a switched-on system running in conservation mode: stress-adapted rather than resilient, under-fuelled at the cellular level despite adequate intake of 'good food', all of which quite clearly explained the brain fog, low mood, water retention, and stubborn inability to regulate weight effortlessly.

That was the moment I stopped guessing and started syncing my habits to actual biology.

Reinvention is not about putting in more effort, but focussing on what matters

What followed wasn't about doing more. It was about doing what mattered, with acute intention. *Radical mindfulness* was introduced earlier in the book—moving it from an abstract idea to a well-oiled operating system created a transformative daily practice: excruciating slowness, protein-first portion calibration, and delayed integration began with a single meal, then every meal, then every waking moment that covered the whole architecture of the day, involving work, sleep, and relationships. Regulating my gut and mindset was the hack to hijacking poor metabolism and strangely, simultaneously informed the manner in which I could relate to everything in life, not just food alone. Regulation, I learned, once made a consistent philosophy in one's day-to-day, makes everything elegantly intuitive from there on out.

This new-found intuition thereby made lifestyle refinements easy, not difficult or forced. I anchored prebiotic foods and protein to more regular meals and switched my workout routine to complement my nervous system, thus reducing stress and inflammation instead of aggravating it. Mixing up strength training, resistance training, daily long walks, and yoga strengthened the mind–body communication more than any hard cardio or high-intensity interval training (HIIT) session I forced myself to do before, and that was made all the more clear with slowed-down eating and a healthier gut, while supercharging my metabolism.

The effects were unmistakeable. I retested everything, this time including my gut health via a stool test. Markers of stress and glucose handling were all optimised and drastically different from what they were before. More importantly, I *felt* like a whole new person. Sleep deepened, the persistent bloat vanished—my body stopped clinging and started trusting. I was able to include all foods without feeling out of control—a marvellous feeling for a lifelong foodie. Vanity steered my efforts to be healthy, but even after ten kilograms dropped off effortlessly, my stomach flattened, and my hair and skin began to shine like they never did before, it was the mental clarity and newfound harmony with my body that truly brought me back to myself. That part felt grounding and gratifying.

My brain felt primed to receive, digest, and most importantly, enjoy. The beauty of this approach of enjoyment and slowness was that it felt more sustainable than anything else tried over the years; changing up diet and exercise did little to resolve bloat and fatigue that ebbed and flowed beneath the surface—rather, returning to refining the internal state, restoring gut function, regulating my nervous system, and aligning my identity with a quieter, more grounded version of myself allowed the external aesthetic outcome to arrive naturally. The result is a body and mind that is easy, not difficult, to work with.

How to use this book as a labour-saving resource

This section of my book is not prescriptive; it is a guiding resource. I severely dislike telling people what to do, but I also understand why people crave instructions—why trust your internal signals or Google skills to know what's best to do first? It is more efficient to outsource this indecision to a clear, science-backed guide.

Self-experimentation and education are essential, since even the most clear science-backed evidence must be altered to fit you and your lifestyle. *Reset* to become aware of what your problems are in the first place, *rewire* to understand the mechanisms, and then here, *reinvent* a new you with one or more of the listed changes. A little later, you'll see a checklist of habits you will benefit from implementing: start with implementing one change and live with it for a week. Do not proceed to add on another habit until this first week is fully complete. Visually crossing out the days in which you successfully enforce a habit, whether that's incorporating a gut-friendly item into one meal or drastically slowing down and spreading a meal out over twenty minutes, is one of the more motivating and effective ways to track progress.

What you don't test and track, you will never properly understand or change. What you observe and try becomes actionable. This is the philosophy I live by, and the one by which I guide others more personally in my coaching too.

After raiding all the literature sources and after experimenting with this template myself, I can tell you this: there is a way to secure the basics, beyond just blood and saliva panels, there is. Once the basics are secured, you can bend the rules (such as during travelling and celebrations) without collapsing and regressing to old habits. Follow the basics, then tailor. It is that simple, and anyone can do it.

Your gut is a personal health fingerprint

The reality is that there's no one-size-fits-all solution for achieving optimised physical and mental health. Just as the keto diet might work wonders for your husband but feels impossible for you to maintain during a raging menstrual cycle, the journey to gut health optimisation is just as varied. What works for your friend Joe may not necessarily work for you, and that's because the differences in our hormonal biology, metabolic responses, and environmental influences make us uniquely individual. As a woman, you would find it especially unhelpful to compare yourself to a man since we are really not little men and have a totally different hormonal cycle and physiology to work around.

But one thing remains clear: it all starts with the gut. Let's take a moment to recap—our guts aren't just another set of organs, like the heart, liver, or lungs (as important as those are, of course). Rather, our guts are fundamental to who we are. During my time at Amili, where we harvested our unique, mammoth Asia-specific gut microbiome database to manufacture products and generate personalised solutions for people, we consistently identified gut-derived metabolites linked to metabolism, digestion, and a wide range of chronic symptoms. In contrast, clinics relying solely on blood panel markers such as C-reactive protein to assess inflammation, or thyroid markers to evaluate metabolic output, were often left with a one-dimensional snapshot of patient health. Our guts being such a core biological mediator means that the health of our guts represents the upstream management that precedes downstream markers such as this CRP. These include not only elevated inflammatory markers, but also visible metabolic markers such as stubborn weight gain, insulin resistance, digestive dysfunction, low mood, anxiety, and ultimately the way individuals experience and move through their lives. In essence, we don't just *have* guts; we *are* our guts. Without their influence, we wouldn't feel, think, or even exist as we do. This is what this chapter is for: to tell you how you can optimise your habits so that your body and mind stop working against you.

That said, there are still significant gaps in research on how gut health affects our physical and mental well-being, and vice versa. For instance, we still don't fully understand the differences between men and women in gut health responses, or the impact of ethnic diversity within a region, be it Southeast Asia, since our diets in Singapore are still quite different from that of someone in Indonesia, or that between London and Norway—environmental and cultural factors may shape these differences. These gaps raise critical questions about the implications of gut health on chronic disease and long-term well-being.

Fortunately, there are remarkable minds and institutions dedicated to tackling these challenges. In Singapore, where gut health research is gaining momentum and public awareness is growing, start-ups are making pioneering advancements in this field. Amili addresses a major issue: the lack of Southeast Asian and broader Asian gut health data. By leveraging cutting-edge technology, they're developing personalised, science-driven solutions that cater to the unique needs of individuals. I worked there and know the management and data scientists, so yes, you can claim there's an element of bias, and for good reason: without region-specific data, much of what we import from the West risks being less relevant than we think, especially since our guts reflect our environment!

There are also science-backed wellness brands such as Wellsprout, which differentiates itself from other greens powders by priding themselves on being gut health-forward, basing their clean ingredient list off over one thousand published studies. It's prebiotic-heavy ingredients do well to nourish the bacteria in your gut and recalibrate the internal cues you need to think, eat and move through the world in a more heightened state of consciousness. Its edge also lies in its versatility, seamlessly blending into everything from olive oil, milk, water, and it can be sprinkled on whatever you're eating as well. This easy and quiet merging of supplementation with real food and real routines and even real pleasure is the future of sustainable wellness, one I'm glad to partake in and steer my future children towards.

Precision intervention: beyond probiotics

I nursed a peculiar fascination with oral administration of faecal transplantation. This as a practice is traditionally administered via colonoscopy, or straight up your butt, to transplant a healthy stool environment (from healthy, screened donors) into a diseased one. Historically, this has been used for *C. difficile* infection, a bacterial infection that specifically infects the large intestine. However, Amili and other places in the world have pioneered the use of oral transplantation, which is exactly what you think it is—swallowing poop in a pill form. Sure, this sounds gross, and for good reason, but both humans and animals have been eating their own stool since the dawn of our very existence. The Chinese have been mixing poop into a slurry, or 'yellow soup', since ancient times to relieve diarrhoea and food poisoning, and some indigenous communities in Australia and South America still practice giving small amounts of stool from healthy community members to treat some acute digestive illnesses.

It is unique and clinically advanced, something which not just harks back to our ancestral yellow soup, but actually has shown clinically significant results in a wide range of medical disorders, including gut and non-gut disorders. To date, there are at least eighty-five clinical indications for the use of faecal transplantation, including IBS, IBD (irritable bowel disease, or a true inflammatory disease, which spans both ulcerative colitis and Crohn's), diabetes, depression, anxiety, ADHD, and ASD (autism spectrum disorder). I pitched to all the doctors and wellness centres I lived with to try as a differentiating factor. This kind of personalised approach represents the future of health—a future where understanding and nurturing the gut isn't just a trend but a cornerstone of how we live, thrive, and heal.

'Just getting a probiotic' isn't the same thing

You might be thinking: why should I go through the effort of spending hundreds of dollars on these personalised tests, supplements, faecal transplants, and reports when I can go and get a high-quality probiotic? Well, that's exactly the problem. Doing that is just like buying a random pair of shoes off the rack without checking if they fit—sure, you might get lucky, but more often than not, it won't be the perfect match for your needs or style.

Obtaining a personalised gut health report is the best way to start, because they help you stop guessing. You can head to any supermarket or wellness store and buy any decent-looking probiotic, but after my personal experience trying to put together a viable recipe for a high-quality, health-optimising probiotic, I realised so much of what you may find in the pharmacy or supermarket may very well be bogus. [43] First, just because a bottle says there's more CFU (meaning colony-forming units, which is the unit of measurement for the millions or billions of bacteria present in a probiotic capsule), doesn't mean it's going to be more effective. Bacteria, like many other organisms, are constantly competing for resources, space, and function within the gut. This is why simply picking a probiotic off the shelf without proper guidance can often lead to suboptimal results. If you don't already take a probiotic, it's worth considering—but it's essential to do it the right way. Think of probiotics as a form of health insurance in a pill, but one that works best when tailored to your specific needs.

That's why personalised gut tests aren't just a fancy extra. They help you stop guessing. Whether you're dealing with anxiety, bloat, brain fog, or all of the above, they help you find the support your body's actually asking for. And that's when healing really begins.

Tracking and adjusting for optimal gut health

Everyone's flow chart for optimised health looks different, and almost every day, there are at least a few new studies I can find on how a certain environmental or biological factor will significantly affect your gut's health and thus your mental and physical well-being.

Take immunity, for example. There have been substantial studies showing the impact of vitamin D on immunity, specifically a combination of vitamin D3 and probiotics. [44] A randomised controlled trial showed that the group that received vitamin D with probiotics experienced a significant increase in mental well-being, alongside reductions in body weight, body mass index, and fat mass. [45] Further, lack of vitamin D has been shown to trigger dysbiosis, which contributes to larger issues such as obesity, type 2 diabetes, and mood disorders such as depression and anxiety. [46, 47]

Nowadays, there are growing and myriad ways in which you can identify problems with your gut and overall health. You can even take a breath test to test for IBS severity. [48] It detects hydrogen and methane concentrations, indicating overgrowth of bacteria that are fermenting unabsorbed carbohydrates in the intestine. [49] IBS is typically comorbid with anxiety and depression, something my old start-up co-founder experienced back when we were trying to make things work.

Back in a 2024 Taiwanese conference, I was fascinated by how the researchers and presenters were mixing ancient Chinese health wisdom with modern Western technology and medicine, and places like Taiwan and Japan are typically at least ten years ahead of the West in terms of technological innovation. There was everything from nail-fold spectroscopy (to analyse blood circulation) to advanced microbiome analysis, wearable health trackers, and integrative approaches combining traditional Chinese medicine (TCM) with cutting-edge biotechnological tools. This seamless blending of old tradition and modernity showcased a future where personalised health solutions could bridge the gap between ancient practices and the scientific advancements of today. The challenge, of course, is not access to information; it is integration. Personalisation breaks down when systems become too complex, when wellness noise is just too overwhelming.

How to deal with derailing obstacles

To make personalisation easier, we can work backwards from the three gut–brain connections introduced in Reset: *neural*, *immune*, and *chemical*, linking root to tail.

Neural: refine gut–brain signalling quality

As discussed earlier in the book, neural gut–brain signalling is largely mediated by the vagus nerve and shaped by the nervous system's ability to accurately perceive internal states. By the time I reached this phase of my own health journey, I no longer needed to understand *what* neural signalling was; I needed to learn how to *refine its quality*.

What became increasingly clear was that most of my symptoms were not the result of ignorance or lack of effort, but of a nervous system that was rarely in a state receptive enough to receive, digest, or interpret signals correctly. Persistent bloating, fatigue, brain fog, or erratic appetite are often framed as personal failures when in reality, they just reflect a nervous system that has not been given the conditions it needs to function coherently. Digestion, appetite, mood, and energy are not governed by discipline alone; they depend on whether the body feels safe enough to engage parasympathetic processes and say, 'OK, I trust you.'

Supporting neural signalling, then, is not about doing more. It is about reducing interference, slowing down meals with radical mindfulness, creating pauses between work and social life, and really feeling present for each moment of the day. Choosing the right forms of movement helps with this: Yoga and Pilates are especially good to incorporate into your existing fitness practice, for aside from being excellent at toning the small muscles you otherwise would not use in your day to day, unlike other exercises which are mostly time and outcome-driven, these are *internal-reference driven* and cannot be done without acute awareness of your posture and linking breath to movement. This type of slow, controlled, continuous movement demands consistent attention to your internal state, thus improving body mapping, or knowing where your body is in space, which clarifies the sensory input to your brain via vagus nerve signalling (also called vagal tone) because your body isn't constantly scanning itself for threats. Your vagus nerve also runs through the chest and abdomen by virtue of being the physical bridge between your gut and brain; therefore, the deeper inhalations and exhalations and engagement of the core mechanically stimulate this nerve to strengthen your parasympathetic, or rest-and-digest, system. Long walks, controlled breathwork, and deliberate stillness are similar tools of signal correction. As interoceptive accuracy improves, the ability to feel hunger, fullness, tension, and ease, or the group of the brain's predictive models, also begin to update, so you can see how simply paying attention to movement and staying present with the world round you has an incredibly significant effect on your signalling quality, which reduces chaotic, inflammatory signalling and thus softens your sense of urgency, helps cravings lose their compulsive edge, and you feel in control and calm always.

Vanity as neural feedback, not superficial reward

The concept of vanity almost earned its own section, but it is so tightly interwoven into neural signalling that isolating it felt almost wrong. Vanity, a philosophy I duly preach, is often dismissed as shallow or self-indulgent, yet in practice it functions as one of the most powerful reinforcement mechanisms available to the nervous system.

When gut–brain signalling begins to regulate, the body reflects that internal order outward. Your posture changes. Facial tension softens. Skin clears, and energy steadies. Body composition shifts without force. All these changes are not incidental; rather, they provide the brain with visible confirmation that internal signals are aligning—that the system is working. It's easy to see in others: someone who looks puffy and swollen all the time is clearly in a stressed-out state or trapped in a lifestyle protocol that is inflaming them from the inside out. The cool thing is also that this is a bidirectional relationship: When the external stuff is attended to intentionally, be it through posture, grooming, or self-presentation, these external signals feed back into your nervous system, reinforcing the internal regulation that made them possible too. This matters more than we like to admit.

For much of my life, mirrors were either avoided or weaponised. They were tools girls in my high school used for criticism and comparison, like that *Mean Girls* episode where all they could do in a mirror together was point out flaws. The brain is a smart thing always listening to you, so if that is all you do when passing a mirror, you rewire your brain to acknowledge poor posture, a flabby belly, and swollen moon-faces as your objective truth, During my own healing process, I deliberately reframed my relationship with my reflection through what is, in effect, a form of mirror exposure therapy. Each morning, post-shower, I would stand in front of the mirror—not scanning for flaws or planning improvements, but simply observing, noticing my posture, how tense I was in the moment, or a sense of ease on easier days. At first, this felt awfully strange, but as time passed, exposure made me motivated to continue working with, and not against, my body. Concurrently, as the work on aligning the internal with the external state progressed, confidence in the streamlined lines and curves of this amazing body, which can do so many things, grew. This bidirectional relationship between inner regulation and outward presentation sits at the core of the Healthy and Vain philosophy.

What changed was not how I looked, but how accurately my brain mapped my body. Regular mirror exposure improves body mapping, which is also enhanced in the long strides, stretches, and toning in yoga or Pilates. As this map became more coherent, eating behaviours softened because I was paying acute attention to my gut and stomach, how it felt during all of the twenty minutes it would take to finish a meal. The urge to micromanage food then diminished, and confidence emerged naturally.

Over time, vanity—expressed through grooming rituals, posture, clothing, and self-presentation—becomes more about neural reassurance than just aesthetics. The facial and scalp massage provided is repeated vagal input. Upright posture enhances vagal tone too, and combining this with long walks supports gastric motility. Circadian-aligned morning routines grounded the nervous system before the day imposed its demands. Each of these acts fed information back to the brain: *This body is regulated; it is safe to digest, to rest, to trust. You can now let go of excess water, fat, and distrust.*

Vanity shaped me and will shape you into a more confident and healthier version of yourself, not because it helps your nervous system learn and inhabit this sense of regulation and stay there. Aligning the internal with the external reinforces behaviours that sustain health in the first place.

Writing as Regulation

On this note of vanity, self-care is often framed as something we do to the body—what we eat, how we move, how we look after our skin. But one of the most overlooked forms of self-care happens on paper, through journalling. I have journalled since the age of 7, and changing journalling prompts and making it a consistent habit during my lifestyle shift has been transformative. The key thing here is that writing is not reflection for reflection's sake, but a form of neural housekeeping.

The brain, like the gut, is simply not designed to hold on to everything. When thoughts, worries, or half-formed emotions are left unprocessed, they occupy bandwidth in your brain, activating the stress response, and over time, they interfere with digestion, appetite regulation, and gut signalling.

To make it super gut health-esque, journalling acts as a form of *cognitive digestion*. It allows experiences to be broken down, named, and metabolised rather than endlessly swirling around in our heads. Our bidirectional gut–brain axis means that a dysregulated nervous system alters gut motility, secretion, and microbial behaviour, just as an inflamed gut feeds back into mood and cognition.

From a physiological standpoint, writing down and thus externalising thoughts reduces cognitive load on the prefrontal cortex and dampens unnecessary sympathetic, or stress, activation. This is why journalling often pairs so well with physical rituals—skincare, massage, posture resets, all of which bring you back into your body and calm the nervous system, creating the conditions for both mental clarity and digestive ease. From a lived standpoint, it creates a sense of containment. Nothing feels as grounding as journalling with a glass of wine and a candle at the end of most nights after shutting off the kitchen and getting in late-night steps with my partner. It is a great transition, signalling to the body that it's time to reflect, rest, and reset before the next day. Importantly, journalling here is not about self-analysis or perfection. It is about making space. Much like regular bowel movements signal a healthy digestive system, regular mental offloading signals a healthy nervous system.

Immune: contain inflammation and restore trust in your body

In addition to a nutrient-rich diet, managing stress, getting adequate quality sleep (at least seven hours, waking up and sleeping at roughly the same times every day), and maintaining regular exercise are essential for immune health. [50, 51] Protecting the gut barrier is a top priority—incorporate bone broth or broths rich in collagen, such as Chinese lotus-root broth, beef broth, or pho, into your diet. These broths help strengthen the gut lining. Make sure to include a regular intake of meat (specifically that rich in omega-3, such as fish), eggs, and L-glutamine supplements as well if you feel your diet would not adequately support this supply of nutrients. What is often overlooked is that immune regulation does not occur in isolation, hidden neatly behind blood markers and lab values, but it also leaves visible traces on the body. The skin, scalp, and oral cavity are immune interfaces, not merely aesthetic surfaces. Persistent acne, eczema, scalp flaking, gum bleeding, facial puffiness, or water retention are often read as cosmetic inconveniences, when in reality they frequently signal low-grade immune activation and barrier stress.

This is where vanity, used intelligently once again, becomes a form of immune care. Gentle, consistent skincare protects the skin barrier and reduces inflammatory signalling driven by irritation or overtreatment. Scalp care matters for the same reason, since chronic scalp inflammation is still inflammation, even if we are socially trained to ignore it. Oral hygiene, often dismissed as routine maintenance, plays a quiet but significant role in immune tone; inflamed gums and dysbiotic oral microbiota contribute to systemic inflammatory load far more than we realise. Even the vanity cues many of us instinctively notice, including morning facial puffiness or what we sometimes call moon-face, abdominal distension, or that heavy, waterlogged feeling, can be reframed as immune feedback rather than flaws to correct. When read this way, vanity stops fuelling criticism and starts becoming hints at what's going on inside us, nudging us towards choices that reduce inflammatory burden instead of accumulating it day by day. In this sense, caring about how you look is not superficial; it is often the most immediate way the immune system asks for gentler inputs.

Back to exercise, briefly. Movement is vital because we are built to move. At a fundamental level, regular physical activity reduces inflammatory load and supports overall gut health. However, the *type* of exercise matters. More is not always better. Zone 2 activity, or where you are exerting yourself but can still comfortably hold a conversation, is particularly supportive for immune regulation. This intensity improves mitochondrial efficiency, supports metabolic health, and lowers chronic inflammation without triggering excessive stress responses.

You can estimate your Zone 2 heart-rate range by first calculating your maximum heart rate (220 minus your age), then taking roughly 60–70 per cent of that number. This is the range to aim for during Zone 2 exercise. Ideally, aim for around forty-five minutes every other day, or even daily if it feels sustainable. Brisk walking, swimming, cycling, and gentle yoga all fall comfortably into this category, so combining the aforementioned yoga and Pilates with a Zone 2 habit is golden. As with most things in health, consistency and restraint matter more than intensity. The goal is to support the immune system, not challenge it unnecessarily.

Chemical: practical ways to optimise gut–brain chemistry

At this stage, the goal is no longer to understand chemical signalling, but to *shape* it deliberately through sustainable lifestyle habits. Gut–brain chemistry responds quickly to inputs: food timing, nutrient availability, microbial support, and stress chemistry, which makes it one of the most malleable pathways to work with.

To support serotonin and GABA production, ensure you are consistently supplying the raw materials rather than sporadically chasing mood fixes. Tryptophan-rich foods such as chicken, turkey, eggs, salmon, and nuts provide the substrate, but they only work when paired with a gut environment capable of converting them efficiently. Regular inclusion of fermented foods and targeted probiotics helps here. Certain fermented foods, such as kimchi and Pu-erh tea, naturally contain GABA, while microbes like *Lactobacillus rhamnosus* support endogenous GABA production. [52] Magnesium is non-negotiable in this context—without it, GABA activity is blunted. Leafy greens, almonds, and cashews are also very reliable food sources.

Short-chain fatty acids (SCFAs) are optimised not by supplements, but by feeding the microbes that produce them. This means eating prebiotic fibres consistently, not occasionally: garlic, onions, leeks, bananas, chicory root (inulin), and resistant starches such as cooked-then-cooled potatoes or rice. Polyphenol-rich foods—berries, green tea, dark chocolate, and yes, red wine—further enhance SCFA production by acting as microbial substrates. The first time I swapped a late-night cookie for a handful of cooled-rice sushi, the steadier energy felt like switching on studio lights in my brain—brighter, calmer, and far less chaotic.

This is also where *vanity becomes chemically useful*. Many of the aesthetic cues people fixate on, such as dull skin, facial puffiness, bloating, or brittle hair with those awful split ends, are downstream reflections of stress chemistry, blood sugar regulation, and bile acid signalling. Chronic cortisol elevation, which happens as a result of chronic stress, alters fluid balance and glucose handling. Disrupted insulin signalling affects fat storage and energy availability. Impaired bile acid metabolism influences digestion, inflammation, and skin clarity. The body often advertises these chemical states outwardly before blood markers fully catch up. When read correctly, vanity stops being a source of pressure and becomes real-time chemical feedback.

It is our responsibility to use this feedback deliberately. Eating in a fed, regulated state stabilises insulin and leptin signalling far more effectively than restriction ever will. Regular meals and predictable rhythms calm ghrelin and reduce reactive hunger. Supporting bile acids—through adequate dietary fat, bitter foods, and microbial diversity—improves digestion and metabolic flow, often showing up quickly as reduced bloating and clearer skin. Even pleasure-oriented self-care, such as warm showers or baths with scented oils, shifts neurochemical balance, lowering stress mediators while supporting dopamine and oxytocin pathways that make consistency possible. Over time, this internal recalibration is reflected outwardly as reduced facial swelling, improved skin quality, and stronger hair, all signs I myself experienced after recalibrating my chemical signalling through vanity and otherwise hinted at a system moving away from stress and towards repair.

Finally, optimise the cofactors that allow this chemistry to run smoothly. Vitamin B6 is essential for serotonin and GABA synthesis [53] and is found in chickpeas, bananas, and potatoes. Vitamin D regulates serotonin synthesis more broadly [54] and should be maintained at sufficient levels through sunlight, food, or supplementation.

Crucially, vanity sustains chemical regulation not through restriction, but through *motivation*. Wanting to feel light, clear-headed, and comfortable in one's body encourages behaviours that stabilise blood sugar, reduce inflammatory metabolites, and support microbial balance—far more reliably than discipline alone. In this way, aesthetic awareness becomes a chemical ally. It keeps the internal environment hospitable, not by forcing outcomes, but by reinforcing the conditions in which healthy chemistry can emerge and remain stable.

Travel

As my dad always said, *if you fail to plan, you plan to fail.* I only truly understood this years later, especially on the road when I started to travel a lot more for both work and play. Travel can throw off your gut rhythm faster than a red-eye flight with no sleep and airport food for dinner. Whether it's a business trip, vacation, or family visit, here's how to stay grounded when everything around you is in motion, the goal not being control, but regulation, just in motion.

The first layer of regulation is thus preparation. Pack a gut kit with supplements tailored to the number of days you will be away from home. It should include targeted probiotics, electrolyte sachets or a greens powder to shake into your water. If it's a longer trip, I'll bring single-serving packets of prebiotic fibre or collagen. These take up barely any space and make a big difference, quietly protecting digestion when your usual routines disintegrate. If needed or if you want to go one step further, pad this little kit with vitamin D, magnesium citrate for bowel regularity and deep sleep, omega-3s, and glutamine for optimal gut protection—not as perfection, but as insurance. You can include this gut kit in your overall wellness kit if you plan to pack resistance bands, Pilates socks, and gym wear to remind yourself to head to the gym at least a few times during the holiday.

The second layer is anchoring. Anchor your day with familiarity and pauses during mealtimes. Keep your morning and evening routines as consistent as possible. That means starting the day with a walk, even if it's just ten minutes around a new block, journalling to record new moments and how you feel around new foods and experiences, and making time to implement radical mindfulness: excruciating slowness, portion calibration with protein, and delayed integration.

The third layer is movement and hydration. Travel is both dehydrating and stagnant. Plane cabins, car rides, and hours of sitting all slow digestion, gut motility, and increase inflammation. It helps to always carry a water bottle and refill it as you go through the day, and set reminders to move: stretches at rest stops, standing up on flights, or even walking the hotel hallway before bed. Movement matters because digestion thrives on motion.

Finally and most importantly, flexibility matters more than control and discipline here. It's very hard to maintain similar routines when you're thrust into a completely new environment—you won't eat perfectly, and jet lag will blur the edges of all the routines and rituals that keep you grounded. But consistency over time matters more than any single day. The ability to meet change without resistance trains the nervous system to remain calm in novelty. In fact, keeping an open and curious mind during periods of change, however drastic, is key here: over time, this rewires your brain towards adaptability rather than threat. And since, as you now know, the gut and brain are in constant conversation, this adaptability is felt physically, not just mentally. What matters is not any single disrupted day, but the pattern you return to. Travel is meant to be lived, explored, and enjoyed, and a resilient gut makes that easier, not harder.

Plan lightly, pack intelligently. Anchor a few non-negotiable things, such as the supplement stack you decide to bring every holiday or road trip, movement, pauses during and after your meals, your habits, and establish your non-negotiable of your supplement factory, making it a point to go to the gym or a Pilates class a few times a week, reduce the stress factors as much as possible. These will really help your body to carry you, no matter the time zone, no matter your environment.

Menstrual cycle

Our menstrual cycle is a big one here. It is not something to manage or override, but something to embrace and work with—I gained a newfound appreciation for this dynamic system that shapes how we think, feel, and move through life after losing mine for a few years during the restrictive period in high school. Every cycle is a reminder of my health, fertility, and a thriving internal ecosystem. Indeed, hormonal shifts across the female cycle do not just affect the uterus. They influence our microbiome, digestion, inflammation, mood, energy, and appetite all at once. Progesterone slows digestion, while changes in oestrogen affect serotonin. The gut registers these fluctuations whether we acknowledge them or not. When you feel bloated, foggy, more emotional, or unusually tired, it is not imagined, and it is not weakness. It is simply your gut responding to a changing internal environment.

This is where radical mindfulness becomes most useful. Rather than forcing the same habits every day, this phase of the work is about noticing, adjusting, and responding honestly. Timing matters. If you practise fasting or more metabolically demanding routines, the follicular phase, roughly the first ten to twelve days of your cycle, is often the most receptive window. Energy tends to be steadier, digestion more efficient, and the body more tolerant of challenge. By contrast, the luteal phase asks for restraint and softness, something I discovered too in my testing and tracking phase. Inflammation naturally rises, while digestion slows. Stress and emotional volatility somehow always followed suit. As progesterone rises, smooth muscle contraction slows, including in the gut. Since transit time lengthens, fermentation increases, and digestion becomes less efficient, I noticed that the same croissant or *popiah* I tolerated effortlessly two weeks earlier would suddenly leave me bloated and uncomfortable, regardless of the meal timing or environment. Practising radical mindfulness here meant resisting the urge to label foods as the problem and instead adjusting *how* I ate—slowing down, always walking after meals, and then some, and accepting that digestion would simply take longer. The discomfort eased not because I restricted more, but because I stopped asking my gut to perform at a speed it was biologically not designed to maintain during this phase.

Why walk? When the body feels tender, heavy, or unsettled, walking becomes one of the most supportive choices available. Not because it is productive, but because it keeps digestion moving gently, lowers stress chemistry, and keeps that mind–body signalling we want to preserve keenly intact. When in doubt, walk. Walk before meals, after meals, first or last thing in the day, if you have time amid life and work demands. Ten minutes is often enough to shift both gut sensation and mental state, especially on days when anything more feels like resistance rather than care.

Cravings deserve the same mindful attention. Across our cycle, appetite and desire shift for valid biochemical reasons. Radical mindfulness here does not mean suppression or indulgence, but presence—returning to the principles discussed earlier, then coming back to the body and asking what is actually being requested. Comfort. Sweetness. Grounding. Rest. Eating slowly, without distraction, allowing satisfaction to register, often resolves far more than rigid avoidance ever could. Over time, the body learns which choices leave it calm and nourished rather than inflamed or depleted. In practice, the principles remain really simple. Walk as often as you can, especially after meals, nourish without urgency. Rest when your body is screaming to be rested. Your gut, your brain, and your hormones are in constant conversation. When you listen carefully, the next step usually becomes obvious.

The same approach applies beyond the menstrual cycle. Travel, social commitments, family life, eating out, and celebrations all introduce variability. While the context changes, the foundations do not. Slow down, eat with awareness and presence, and respect portioning with protein priming. Reinvention is not about rigid control. It is about knowing how to return to balance, again and again, across different seasons of life.

Food Choices Cheat Sheet

In the context of eating out, travel, and everyday decision-making, this section is intended as a practical anchor, not a new set of rules. Think of it as a short, functional cheat sheet rather than a comprehensive dietary guide. Many of the food groups mentioned here will feel familiar, as they intentionally overlap with the Rewire tier, where omega-3s, polyphenols, fermented foods, and healthy fats are introduced in more detail.

The difference lies in *how* these foods are used. In this Reinvent phase, they are not presented as permanent prescriptions, but as temporary supports you can lean on for a defined period of time to recalibrate gut–brain signalling. This is particularly useful during moments of disruption—travel, hormonal shifts, stress-heavy weeks, or when your body simply feels noisy, lazy, and uncooperative.

Rather than expanding the list endlessly, the aim here is simplicity and repeatability. These foods are easy to find, easy to combine, and really forgiving when life gets busy. Used consistently for a few weeks, they help restore rhythm, reduce inflammatory and metabolic friction, and bring internal signals back into focus. Once that baseline is re-established, flexibility becomes far easier to maintain without slipping back into guesswork or extremes.

1. **Serotonin Precursors:** Increase your serotonin levels by consuming foods high in tryptophan:

    - chicken
    - turkey
    - eggs
    - salmon
    - nuts

2. **GABA Support**

    - Include fermented foods such as kimchi or Pu-erh tea
    - Probiotics for GABA-producing bacteria: *L. rhamnosus* supplements or fermented dairy like yoghurt, if you are not lactose-intolerant
    - Ensure adequate magnesium intake: leafy greens, cashews and almonds

3. SCFA Production

- o Prebiotics to fuel SCFA production, garlic, onions, banana, Leeks or chicory root (inulin). Incorporate cooked-and-cooled potatoes or rice (I typically cook a large batch and leave this in the fridge to cool overnight, and you can take out servings if and when you desire).
- o Polyphenol-rich foods for SCFA metabolism: berries, green tea, dark chocolate and red wine

4. Vitamin B6 (Pyridoxine)

- o Essential for serotonin and GABA synthesis, putting this in a positive feedback loop with my first two points
- o Sources: chickpeas, bananas and potatoes

5. Vitamin D

- o Regulates serotonin synthesis
- o Sources: sunlight, fatty fish like salmon, mackerel and sardines, animal liver, egg yolks
- o Supplements: Necessary even in a warmer climate; most people do not get as much vitamin D as they think; aim for 600–800 IU daily (or per your doctor's advice)

Vitamin D is one of those underrated supplements, and most people who live in sunny countries like me in Singapore think that supplementation with vitamin D isn't necessary, for don't we get enough of it from the sun anyway? The answer is a solid no. Vitamin D is needed for hormone synthesis, commonly lacking in those with mood disorders such as depression and anxiety, even, for they lack the building blocks of the very substances needed to make them feel happy and stable.

Keep this cheat sheet in your journal or phone's notes, ready to whip out at the grocery store. Each bullet is a small *reinvent* insight you can deploy on the fly, turning everyday meals and moments into catalysts for steadier mood, stronger immunity, and a gut that's finally on your side.

This beautiful and unique interplay between diet, lifestyle, and targeted interventions underscores the critical role of the gut in shaping our overall health. By taking a tailored approach, whether through advanced microbiome analysis, nutrient optimisation, or easy lifestyle strategies, all combined with radical mindfulness, you are not just improving your health today but investing in a bright future full of resilience and vitality. The journey may be complex, but with the right tools and knowledge, it's one worth embarking on.

Understanding your gut really is understanding yourself.

Before pulling everything together, it helps to understand the rhythm of this process. Gut health is not something you optimise in one step; it unfolds in phases. The first phase is orientation: using data to understand where you are and what patterns might be driving symptoms. The second is regulation: a reset of a few weeks (usually 4-6 weeks) where daily inputs are simplified so the gut–brain axis has space to recalibrate. Only then does refinement make sense. Testing provides a snapshot, not a verdict; the reset provides movement. Together, they turn information into insight rather than overwhelm.

Pulling everything together, here is a comprehensive checklist to ensure you're testing and tracking the right things, at the right times:

Pulling It Together: Test and Track

Test and Track: Baseline Orientation Tools

Start being driven by data

- ☐ **Comprehensive and targeted gut microbiome test** (with diversity, keystone strains, and functional outputs if available)

- ☐ **Urinary neurotransmitter metabolites panel** (e.g., 5-HIAA for serotonin metabolism—optional but insightful)

- ☐ **Blood panel** (including fasting insulin, HbA1C, vitamins D, B12, folate, CRP, lipid profile)

- ☐ **Stool calprotectin and zonulin** (tells you about gut permeability or inflammation)

- ☐ **Food sensitivity screening** (optional, not a must unless you're experiencing symptoms)

Data alone, of course, does not change physiology. What shifts the gut–brain axis day to day are the inputs you repeat—how you eat, move, rest, manage your stress and emotions. The next six weeks are about simplifying those inputs so the system can settle and become more intelligible.

Daily Gut Rituals

Elevate your baseline gut health

- ☐ **Radical mindfulness:** If there is any key takeaway from this whole book, make it this one. As explained above, incorporate excruciating slowness, portion calibration with protein priming, and delayed integration into every single meal. Taste fully, chew thoroughly, enjoy, and be grateful for what is in front of you. If you're in company, focus on the conversation just as much as what you're putting into your mouth.

- ☐ **Include foods from the list mentioned in the Rewire and Reinvent tiers:** specifically fermented foods 1–2× daily (e.g. kefir, kimchi, miso, yogurt), omega-3, polyphenols, good fats, and glutamine. Have fun experimenting with a few of those ingredients for your at-home weeknight dinners.

- ☐ **Hydrate thoughtfully:** start your day with water, and stay lightly hydrated throughout to nourish your gut and stop mistaking hunger for hydration.

- ☐ **Move with intent:** choose a mixture of exercises without confining it to, but try to incorporate yoga and/or Pilates for mind–body connection, and always walk at least a few thousand steps after every meal.

- ☐ **Rest your gut nightly:** twelve to fourteen-hour overnight fast when possible; this allows digestion to complete, insulin to settle, and overnight repair processes to take priority.

- ☐ **Minimise ultra-processed foods:** chips, sweets, anything in some kind of packaging, to protect your precious gut lining.

- ☐ **Supplement smartly and simply:** a high-quality probiotic, omega-3s, vitamin D, magnesium, and glutamine form the foundation, supporting gut barrier integrity, inflammatory tone, circadian rhythm, and nervous-system regulation. If you are more experimental, additions such as creatine, collagen, curcumin, L-theanine, ashwagandha, NMN or SIRT-6, and quercetin (also found naturally in apple skins) may support cellular energy, stress modulation, and repair — but only once the system is already calm, receptive, and regulated.

- ☐ **Limit alcohol:** Champagne counts as soul-filling, so make it a real treat, not a nightly regular. If you are still in the first six-week Reset phase, try not to drink at all or have tasting substitutes or flavoured sparkling water on hand as a substitute.

By now, it should be clear that digestion does not happen in isolation. The gut responds to the nervous system just as much as it does to nutrients. Without addressing cognitive load, emotional processing, and daily stimulation, even the best diet remains partially inaccessible. And now finally we get to . . .

Mental and Emotional Hygiene

The overlooked side of gut health that has nothing to do with food

- ☐ **Journalling through the day:** This can be combined with the Delayed Integration segment of Radical Mindfulness. Some good journal prompts to check in with your mood, dietary habits, and energy are 'What meals made me feel grounded? What am I no longer willing to negotiate about myself? How did my appetite change after eating at this particular restaurant? What changed in my digestion or energy? What sensation lingered in my chest after that rattling argument?'

- ☐ **Protect your mornings and evenings:** Avoid screens for 20–30 minutes upon waking, and again before you sleep. This reduces cortisol dysregulation, which shapes mood, appetite, and the body's tendency to cling to weight. At night, instil an evening routine that feels unhurried and intentional—warm light, slower touch, skincare, a stretch, a few lines on paper—so the body understands it is safe to soften.

- ☐ **Curate your inputs:** Overstimulation dysregulates the gut–brain axis as reliably as poor food does. Engage in podcasts, books, and conversations that stretch your mind while regulating your nervous system (yes, that includes conversations with uplifting, not toxic friends), but

- ☐ **Laugh. A lot.** Even at your own expense. Laughing downshifts the body out of threat, resoftening cortisol output while stimulating vagal tone due to the engagement of the diaphragm and chest, which the vagus nerve runs through—all of which directly influence digestion, inflammation, and mood. A body that laughs regularly is a body that feels safe, not just happy.

- ☐ **Practice mirror exposure.** Not for critique, but for neutrality. Looking at yourself and embracing the reflection regularly helps you learn to meet your reflection without judgement, reduces stress reactivity. Over time, this changes how you inhabit your body and how others read your presence.

- ☐ **The intelligent use of vanity and self-care.** Alongside mirror exposure, incorporate scalp and facial massage into your regular skincare or body routine as a way of returning to the body. Posture reset with a straight back, shoulders back, unclenched jaw; change breathing patterns and appetite signalling. How you hold yourself influences how you eat and how you are read. Even applying body lotion and getting a manicure can pull attention out of the world around you and back into the body. When your sense of attractiveness is regulated, it supports posture, confidence, appetite regulation, and social ease.

Finishing Off: Reinvention as a Way of Living

As I reach the end of this chapter and of this book, I am surprised by how emotional I feel writing these final lines. That in itself feels like proof of the work. There was a time when signals from my body and gut arrived garbled, when emotion felt either overwhelming or distant, or as my therapist put it the very first time we met all those years ago, 'You can't tell emotions apart, can you?' To feel clarity now, not just intellectually but viscerally and deep in my, well, gut, is a reminder of how far this journey has stretched and how much of it has been lived rather than just theorised.

This book has never just been about fixing a broken body, though it did kind of start that way with my warped relationship with my internal signals and external perception of myself, influencing the way I moved through my teenage years. It has been about learning to listen properly to oneself again. When gut health, metabolic health, and mindset are aligned, the body stops feeling like a problem to solve. It becomes a responsive, intelligent system that will work wonders for you, effortlessly aligning the internal state with the external. Goodbye puffy cheeks, dull skin and hair, reactive mood swings, and relationships that continue to suffer; hello effortless leanness, a bright heart and mind, good digestion, high energy, and stable mood. The constant low-grade tension of trying to control outcomes gives way to something quieter and far more powerful: *bodily trust*.

Gut health sets the foundation, and from that centre, a ripple spreads, influencing everything that follows. It shapes how we digest food, experience emotion, resolve inflammation, and generate the chemical messengers that influence mood and motivation. Metabolic health gives this system direction and stability, determining whether the body feels safe enough to release, adapt, and recover. Lastly, our mindset shapes the relationship we have with these signals. When these three pillars come together, the noise fades. Decisions around food and movement no longer feel charged, and the temple that is our body begins to feel like an ally rather than an obstacle.

The Reset phase invited awareness and education, a recognition that many struggles were not personal failures but biological mismatches. Rewire brought understanding, showing why the body behaved as it did and how the gut, brain, and metabolism were always in conversation. Reinvent is where this understanding becomes lived experience, where habits settle naturally because they are aligned with biology, and where effort is replaced by rhythm.

What moves me most, looking back, is realising that what once felt impossibly complex eventually revealed itself to be surprisingly elegant. Years of experimentation, confusion, restriction, and curiosity were distilled into a clear and workable framework, a neat and tight formula that respects complexity without being totally paralysed by it.

As this chapter finally comes to a close, take a moment to reflect on your own shifts. How is your self-image changing as you align more closely with your biology? What happens when control gives way to trust, when you stop chasing outcomes and start creating the conditions that reliably produce them? From here, you can move forward with a different standard, one grounded in self-knowledge rather than self-criticism, guided by internal signals rather than external noise. This is the beginning of a new relationship with food, with your gut, and with yourself. It's a whole new identity.

Kundera himself said in his book *The Unbearable Lightness of Being* that 'when we ignore the body, we are more easily victimized by it'. Everything in this book has been an answer to that sentence, which I kept coming back to while putting this together. So much of what I went through and what I see in others comes from being cut off from the body and trying to manage life purely from the head. *Gutsy* is about understanding that feeling grounded and feeling attractive are not shallow goals, but signs of regulation. If there is one thing I hope you take from this, it is that change does not come from forcing yourself into a new shape. It comes from returning to yourself in the most practical ways: how you eat, how you move, and how you take care of your body when no one is watching—slowly, from the inside out, and in a way that is entirely your own.

ACKNOWLEDGEMENTS

I would like to thank God first and foremost, for re-entering my life when I returned to Singapore and for being the quiet, steady guide behind everything I do.

I want to thank Gus, without whom I might never have found the courage to follow my creative direction or gather the loose, restless thoughts that once hovered around me day after day. I love you so much.

I'd like to thank my colleagues and the network of doctors, academics, and entrepreneurs I became friends with through Amili, including Dr Sharad, Nick, Jon San, Evan, Anna, Dr Jeremy, Dr Arun, Yongwei, and Dr David—whose encouragement pushed me to flourish and pursue this with confidence. I also would like to thank my therapist and prominent academic Dr Gipps, whom I met during my lowest points in the United Kingdom and whom I still see regularly. He has seen me through everything.

Lastly, thanks to my family—Mum, Dad, and closest friends scattered across the world: Anastasia, Leia, Claire, Ruru, Miriam, Aneka, Sarayu, Carla, Katie, Charlie, Lauren, thank you for your unwavering support, for lifting me up regardless of time, distance, or circumstance.

REFERENCES

1. Roohi, E., N. Jaafari, and F. J. J. o. n. Hashemian, On inflammatory hypothesis of depression: what is the role of IL-6 in the middle of the chaos? *Journal of Neuroinflammation*, 2021. **18**: 1–15.
2. Cani, P. D. J. G., Human gut microbiome: hopes, threats and promises. *Gut BMJ Journals*, 2018. **67**(9): 1716–1725.
3. Petrut, S.-M., et al., Gut over mind: exploring the powerful gut–brain axis. *Nutrients*, 2025. **17**(5): 842.
4. Aaronson, S.T., et al., Single-dose synthetic psilocybin with psychotherapy for treatment-resistant bipolar type II major depressive episodes: a nonrandomized open-label trial. *JAMA Psychiatry*, 2024. **81**(6): 555–562.
5. Liu, J., et al., *Ameliorating effects of transcutaneous auricular vagus nerve stimulation on a mouse model of constipation-predominant irritable bowel syndrome.* Neurobiology of Disease, 2024. **193**: p. 106440.
6. Sahn, B., et al., *Transcutaneous auricular vagus nerve stimulation attenuates inflammatory bowel disease in children: a proof-of-concept clinical trial.* Bioelectronic Medicine, 2023. **9**(1): p. 23.
7. Ashique, S., et al., *Gut-brain axis: A cutting-edge approach to target neurological disorders and potential synbiotic application.* Heliyon, 2024. **10**(13).
8. Randeni, N. and B.J.I.J.o.M.S. Xu, *Critical review of the cross-links between dietary components, the gut microbiome, and depression.* International Journal of Molecular Sciences, 2025. **26**(2): p. 614.

9. Pereira-Figueiredo, I. and E.H.J.E. Umeoka, *Stress: Influences and Determinants of Psychopathology.* Encyclopedia, 2024. **4**(2): p. 1026–1043.
10. Xie, M., et al., *Associations between trait and state perceived stress and daily moods: COVID-19 stressful experiences as a moderator.* Current Psychology, 2024. **43**(16): p. 14894–14908.
11. Glotzbach, E., et al., *Prefrontal brain activation during emotional processing: a functional near infrared spectroscopy study (fNIRS).* The Open Neuroimaging Journal, 2011. **5**: p. 33.
12. Šimić, G., et al., *Understanding emotions: origins and roles of the amygdala.* Biomolecules, 2021. **11**(6): p. 823.
13. Klawonn, A.M. and R.C. Malenka. *Nucleus accumbens modulation in reward and aversion.* in *Cold Spring Harbor symposia on quantitative biology.* 2018. Cold Spring Harbor Laboratory Press.
14. García-Cabezas, M.Á. and H.J.C.C. Barbas, *Anterior cingulate pathways may affect emotions through orbitofrontal cortex.* Cerebral Cortex, 2017. **27**(10): p. 4891–4910.
15. Sheng, J.A., et al., *The hypothalamic-pituitary-adrenal axis: development, programming actions of hormones, and maternal-fetal interactions.* Frontiers in Behavioral Neuroscience, 2021. **14**: p. 601939.
16. Herman, J.P., et al., *Regulation of the hypothalamic-pituitary-adrenocortical stress response.* Comprehensive Physiology, 2016. **6**(2): p. 603.
17. Gold, P.W.J.M.p., *The organization of the stress system and its dysregulation in depressive illness.* Molecular Psychiatry, 2015. **20**(1): p. 32–47.
18. Bektaş, A., et al., *Does Seratonin in the intestines make you happy?* The Turkish Journal of Gastroenterology, 2020. **31**(10): p. 721.
19. Handy, A.B., et al., *Psychiatric symptoms across the menstrual cycle in adult women: a comprehensive review.* Harvard Review of Psychiatry, 2022. **30**(2): p. 100–117.

20. Awasthi, S.S.J.C., *Correlation Between Perceived Stress Scores and Menstrual Characteristics in Young Indian Women.* Cureus, 2025. **17**(4).
21. Gasmi, A., et al., *Neurotransmitters regulation and food intake: The role of dietary sources in neurotransmission.* Molecules, 2022. **28**(1): p. 210.
22. Koh, A., et al., *From dietary fiber to host physiology: short-chain fatty acids as key bacterial metabolites.* Cell, 2016. **165**(6): p. 1332–1345.
23. O'Riordan, K.J., et al., *Short chain fatty acids: Microbial metabolites for gut-brain axis signalling.* Molecular Cellular Endocrinology, 2022. **546**: p. 111572.
24. Luan, X., et al., *Abnormalities of lipid metabolism in the progression and treatment of depression.* Frontiers in Psychiatry. **16**: p. 1589663.
25. Saleem, G.N., et al., *Therapeutic potential of popular fermented dairy products and its benefits on human health.* Frontiers in Nutrition, 2024. **11**: p. 1328620.
26. Liu, H., et al., *Dietary polyphenols regulate appetite mechanism via gut-brain axis and gut homeostasis.* Food Chemistry, 2024. **446**: p. 138739.
27. Pal, M.M.J.F.i.H.N., *Glutamate: The master neurotransmitter and its implications in chronic stress and mood disorders.* Frontiers in Human Neuroscience, 2021. **15**: p. 722323.
28. Zuhl, M., et al., *Glutamine Supplementation: A Possible Strategy to Help Mitigate Health Risks of Heat-Related Illness.* Journal of Nutrition Metabolism, 2024. **2024**(1): p. 1638244.
29. Reilly, D.M., J.J.P. Lozano, and A. Research, *Skin collagen through the lifestages: Importance for skin health and beauty.* Plastic Aesthetic Research, 2021. **8**: p. N/A-N/A.
30. Mulligan, E.M., et al., *Effects of menstrual cycle phase on electrocortical response to reward and depressive symptoms in women.* Psychophysiology, 2018. **55**(12): p. e13268.
31. Han, Y., et al., *Vagus nerve and underlying impact on the gut microbiota-brain axis in behavior and neurodegenerative diseases.* Journal of Inflammation Research, 2022: p. 6213–6230.

32. Costa, L.N.G., et al., *Metabolomics of major depressive disorder: a systematic review of clinical studies.* Cureus, 2022. **14**(3).
33. Cryan, J.F., et al., *The microbiota-gut-brain axis.* Physiological Reviews, 2019.
34. Agus, A., et al., *Gut microbiota regulation of tryptophan metabolism in health and disease.* Cell Host Microbe, 2018. **23**(6): p. 716–724.
35. Canfora, E.E., et al., *Gut microbial metabolites in obesity, NAFLD and T2DM.* Nature Reviews Endocrinology, 2019. **15**(5): p. 261–273.
36. Zhang, Y., et al., *Metabolite interactions between host and microbiota during health and disease: Which feeds the other?* Biomedicine Pharmacotherapy, 2023. **160**: p. 114295.
37. Kearns, R.J.I., *The kynurenine pathway in gut permeability and inflammation.* Inflammation, 2024: p. 1–15.
38. Bosi, A., et al., *Tryptophan metabolites along the microbiota-gut-brain axis: an interkingdom communication system influencing the gut in health and disease.* International Journal of Tryptophan Research, 2020. **13**: p. 1178646920928984.
39. Terry, N. and K.G.J.G.P. Margolis, *Serotonergic mechanisms regulating the GI tract: experimental evidence and therapeutic relevance.* Handbook of experimental pharmacology, 2017. **239**: p. 319–342.
40. Barandouzi, Z.A., et al., *Associations of neurotransmitters and the gut microbiome with emotional distress in mixed type of irritable bowel syndrome.* Scientific Reports, 2022. **12**(1): p. 1648.
41. Banskota, S., J.-E. Ghia, and W.I.J.B. Khan, *Serotonin in the gut: Blessing or a curse.* Biochimie, 2019. **161**: p. 56–64.
42. Grondin, J.A. and W.I.J.J.o.t.C.A.o.G. Khan, *Emerging roles of gut serotonin in regulation of immune response, microbiota composition and intestinal inflammation.* Journal of the Canadian Association of Gastroenterology, 2024. **7**(1): p. 88–96.
43. Wolever, T.M.J.T.A.J.o.C.N., *Personalized nutrition by prediction of glycemic responses: garbage in→ garbage out.* The American Journal of Clinical Nutrition, 2025. **121**(1): p. 1–2.
44. Huang, F.-C. and S.-C.J.M. Huang, *The cooperation of Bifidobacterium longum and active vitamin D3 on innate immunity in Salmonella*

colitis mice via vitamin D receptor. Microorganisms, 2021. **9**(9): p. 1804.

45. AlQutub, R., R. El Asmar, and D. Papandreou, *Vitamin D, Immunity, and Gut Health*, in *Vitamin DA Novel Therapy for Chronic Diseases?* 2024, Bentham Science Publishers. p. 16–32.
46. Street, M.E., et al., *The Role of Nutraceuticals and Probiotics in Addition to Lifestyle Intervention in the Management of Childhood Obesity—Part 2: Comorbidities.* Nutrients, 2025. **17**(9): p. 1487.
47. Sukik, A., J. Alalwani, and V. Ganji, *Vitamin D, Gut Microbiota, and Cardiometabolic Diseases—A Possible Three-Way Axis.* International Journal of Molecular Sciences, 2023. **24**(2): p. 940.
48. Rana, S.V. and A.J.W.J.o.G.W. Malik, *Breath tests and irritable bowel syndrome.* World Journal of Gastroenterology, 2014. **20**(24): p. 7587.
49. Onana Ndong, P., et al., *Prevalence of small intestinal bacterial overgrowth in irritable bowel syndrome (IBS): Correlating H2 or CH4 production with severity of IBS.* JGH Open, 2023. **7**(4): p. 311–320.
50. Monda, V., et al., *Exercise modifies the gut microbiota with positive health effects.* Oxidative Medicine Cellular Longevity. **2017**(1): p. 3831972.
51. Nieman, D.C., L.M.J.J.o.s. Wentz, and h. science, *The compelling link between physical activity and the body's defense system.* Journal of Sport Health Science, 2019. **8**(3): p. 201–217.
52. Moore, J.F., et al., *Formation of γ-aminobutyric acid (GABA) during the natural lactic acid fermentation of cucumber.* Journal of Food Composition Analysis, 2021. **96**: p. 103711.
53. Field, D.T., et al., *High-dose Vitamin B6 supplementation reduces anxiety and strengthens visual surround suppression.* Human Psychopharmacology Clinical Experimental, 2022. **37**(6): p. e2852.
54. Yu, X.-L., C.-P. Li, and L.-P.J.F.i.P. He, *Vitamin D may alleviate irritable bowel syndrome by modulating serotonin synthesis: A hypothesis based on recent literature.* Frontiers in Physiology, 2023. **14**: p. 1152958.

Printed and bound by CPI Group (UK) Ltd, Croydon, CR0 4YY
08/05/2026
02106543-0002